oxford **maths**
for australian schools 2

contents

NUMBER AND ALGEBRA

MEASUREMENT AND SPACE

STATISTICS AND PROBABILITY

Practice

1 How many?

a

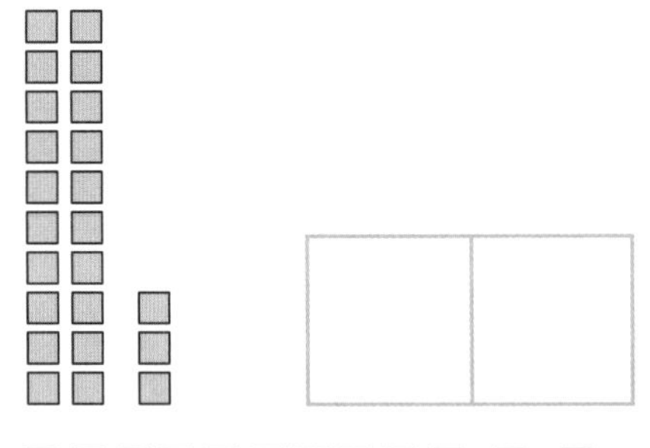

b

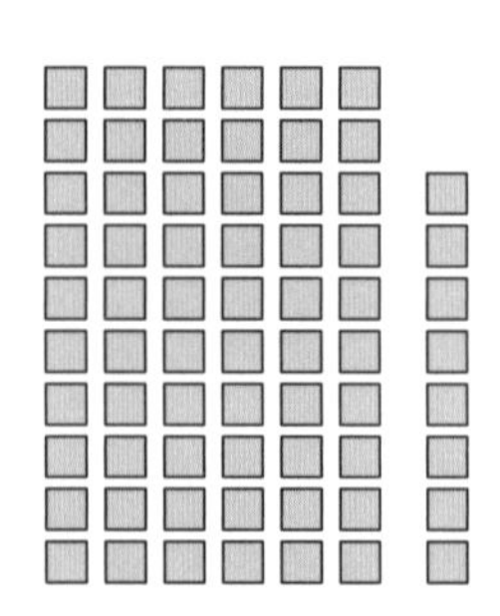

c

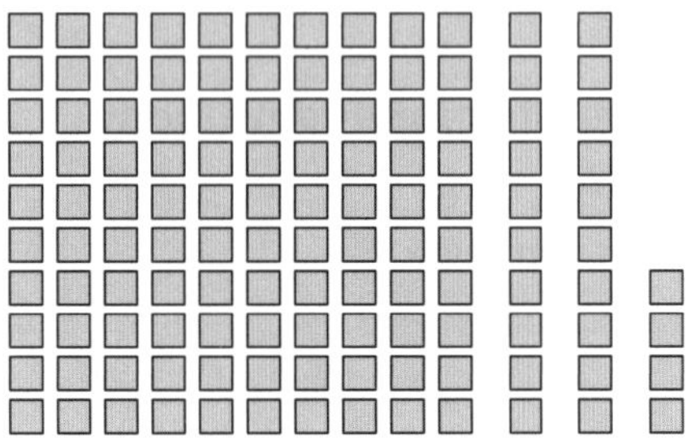

d

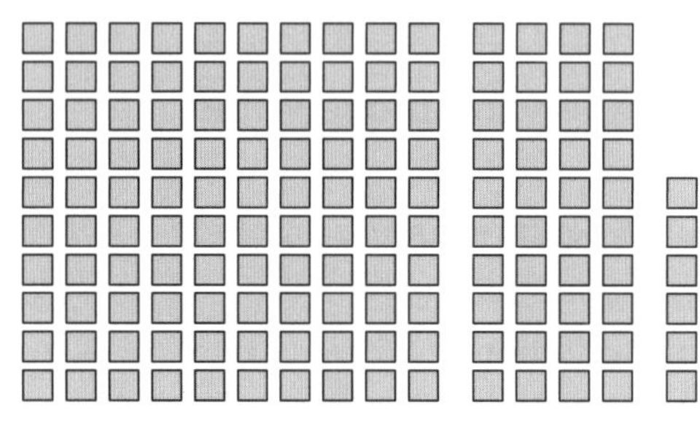

2 Write the numbers.

a

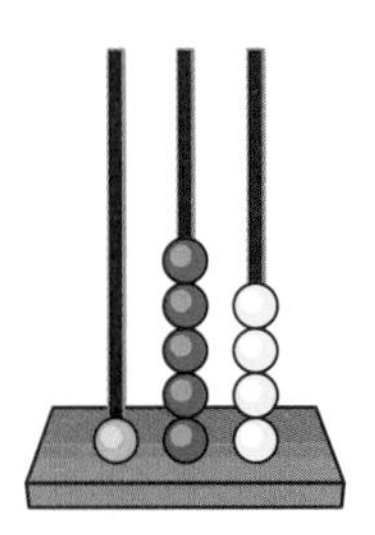

b

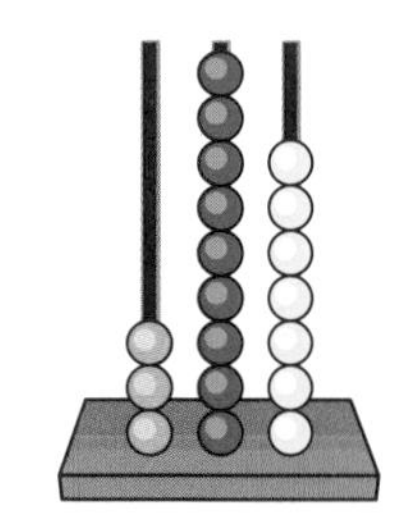

c

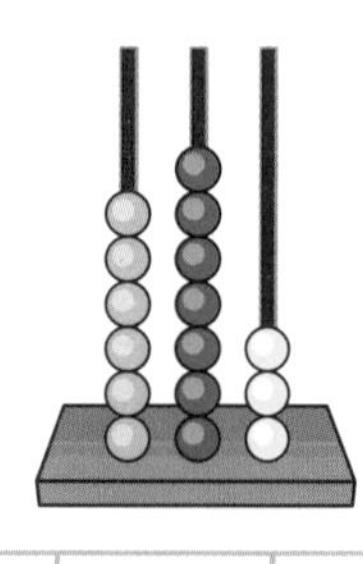

3 This is 219.

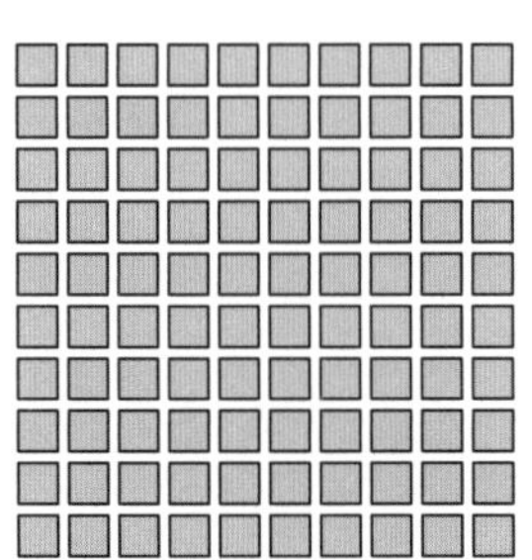

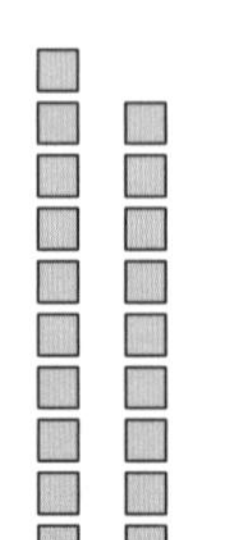

☐ hundreds

☐ tens

☐ ones

Challenge

1

a In 375, how many:

hundreds? ☐

tens? ☐

ones? ☐

b In 607, how many:

hundreds? ☐

tens? ☐

ones? ☐

2

a In 374, the 7 is worth (circle one):

700 70 7

b In 837, the 8 is worth (circle one):

800 80 8

3

Use the digits to make:

a the smallest number.

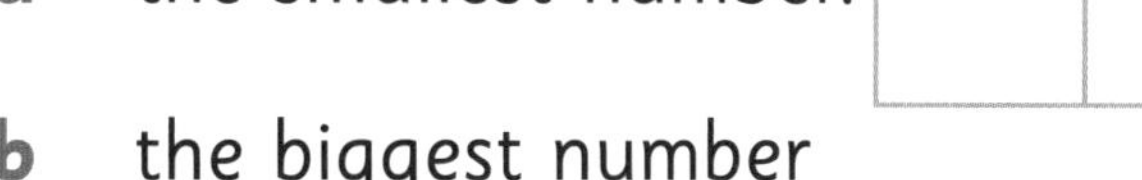

b the biggest number with the 3 in the tens place.

4 Write a different digit in each box.

Use the digits to make:

a the smallest number.

b the biggest number.

Mastery

1 These are parts of a hundred chart.

a Write the number 44 in a square on the third row. Fill in the blanks.

b Write the number 44 in a different square on the grid. Fill in the blanks.

2 Put these numbers in order from smallest to biggest.

132 231 213 123

3 Use each digit once to make as many three-digit numbers as you can.

6 7 8

4 **a** Circle the number that is the best estimate for how many items there are.

3

35

350

3500

b How do you know?

Regrouping and renaming numbers

Practice

Write how the number has been regrouped.

1 671

a

☐ hundreds

☐ tens

☐ ones

b

☐ hundreds

☐ tens

☐ ones

c

☐ hundreds

☐ tens

☐ ones

2 Draw and write a different way to regroup each number.

a 408

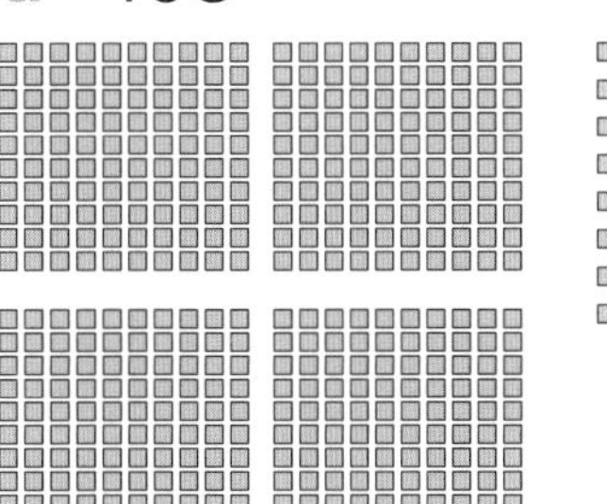

b 316

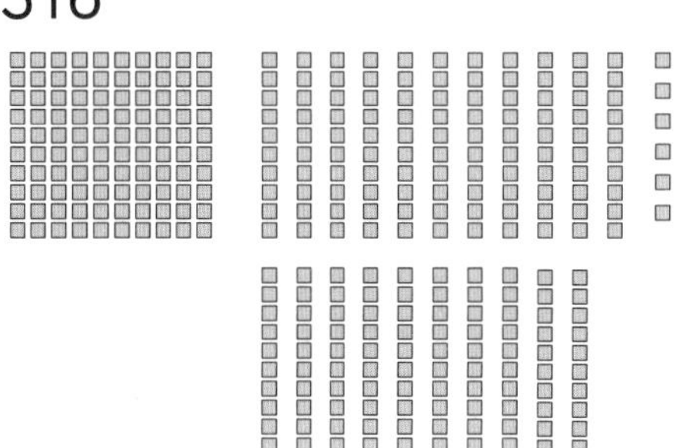

Challenge

1 Draw lines to match the numbers that are the same. The first one has been done for you.

5 hundreds, 14 tens and 7 ones	6 hundreds, 10 tens and 9 ones
3 hundreds and 48 ones	3 hundreds, 12 tens and 54 ones
7 hundreds and 9 ones	6 hundreds, 3 tens and 17 ones
3 hundreds, 15 tens and 8 ones	3 hundreds, 4 tens and 8 ones
4 hundreds, 2 tens and 54 ones	4 hundreds and 58 ones

2 Rename each number without using hundreds.

a 298 ______________________

b 501 ______________________

c 888 ______________________

3 Rename each number without using tens.

a 298 ______________________

b 501 ______________________

c 888 ______________________

Mastery

1 Here are some numbers in Roman numerals:

Roman numeral chart

1	I	11	XI	50	L
2	II	12	XII	100	C
3	III	13	XIII	500	D
4	IV	14	XIV	1000	M
5	V	15	XV		
6	VI	16	XVI		
7	VII	17	XVII		
8	VIII	18	XVIII		
9	IX	19	XIX		
10	X	20	XX		

MM = 2000 DCIX = 609
CDIX = 409 CCLXI = 261

Use the Roman numeral chart to help you write:

a 301 ____________ b 1200 ____________ c 758 ____________

d 525 ____________ e 814 ____________ f 167 ____________

2 The number system we use is called the base-10 place value system. Do you find it easier to use Roman numerals or base-10 to make numbers? Why?

____

____

____

3 The number 7 on Olivia's keyboard wasn't working. Help her by renaming each of the numbers below without using a 7.

a 734 ____

b 576 ____

c 772 ____

Adding in your head

Practice

1 Add these doubles.

a 9 + 9 = ☐☐ b 11 + 11 = ☐☐

c 13 + 13 = ☐☐ d 21 + 21 = ☐☐

e 30 + 30 = ☐☐ f 44 + 44 = ☐☐

2 Add these near doubles.

a 10 + 11 = 10 + 10 + ☐ = ☐☐ + ☐ = ☐☐

b 12 + 13 = ☐☐ + ☐☐ + ☐ = ☐☐ + ☐ = ☐☐

c 15 + 16 = ☐☐ + ☐☐ + ☐ = ☐☐ + ☐ = ☐☐

d 22 + 23 = ☐☐ + ☐☐ + ☐ = ☐☐ + ☐ = ☐☐

e 41 + 44 = ☐☐ + ☐☐ + ☐ = ☐☐ + ☐ = ☐☐

3 Solve by bridging to a 10.

46 + 8 = ☐☐ + ☐ + ☐

= ☐☐ + ☐

= ☐☐

Challenge

1 Use doubles to help you solve these.

a 30 + 32 = ☐☐ **b** 42 + 45 = ☐☐

c 21 + 23 = ☐☐ **d** 34 + 35 = ☐☐

e 49 + 51 = ☐☐☐ **f** 54 + 56 = ☐☐☐

---

2 Use bridging to a 10 to solve these.

a 37 + 7 = ☐☐ **b** 52 + 9 = ☐☐

c 44 + 8 = ☐☐ **d** 65 + 9 = ☐☐

e 95 + 6 = ☐☐☐ **f** 97 + 6 = ☐☐☐

---

3 15 + 16 = 31. Write different numbers in the boxes to add up to 31.

a ☐☐ + ☐☐ = 31 **b** ☐☐ + ☐☐ = 31

c ☐☐ + ☐☐ = 31 **d** ☐☐ + ☐☐ = 31

---

4 Eva buys two books. Write an addition sentence to show how much she spends.

Mastery

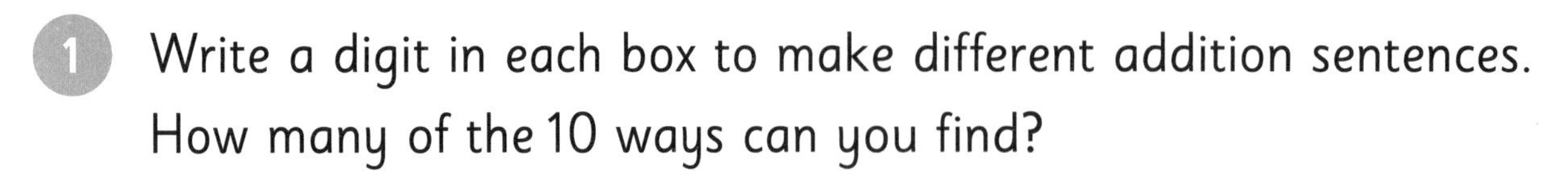

1 Write a digit in each box to make different addition sentences. How many of the 10 ways can you find?

a	4		+	1		=		2
b	4		+	1		=		2
c	4		+	1		=		2
d	4		+	1		=		2
e	4		+	1		=		2
f	4		+	1		=		2
g	4		+	1		=		2
h	4		+	1		=		2
i	4		+	1		=		2
j	4		+	1		=		2

2 Write a digit in each box to make different addition sentences. How many different ways can you find?

a	4		+	4		=		0
b	4		+	4		=		0
c	4		+	4		=		0
d	4		+	4		=		0
e	4		+	4		=		0
f	4		+	4		=		0
g	4		+	4		=		0
h	4		+	4		=		0
i	4		+	4		=		0
j	4		+	4		=		0

3 What if you change the digits? Are there always 10 different ways to make the addition sentences? You may need to use extra paper.

Exploring addition

Practice

1 Use the number lines to help you add.

a

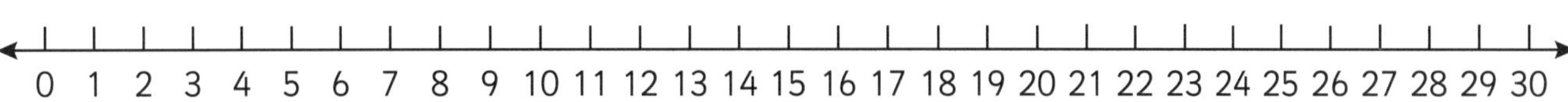

9 + 17 = ☐☐

b

11 + 34 = ☐☐

---

2 Draw and solve.

a 5 + 9 + 1 = ☐ + ☐ + ☐ = ☐☐

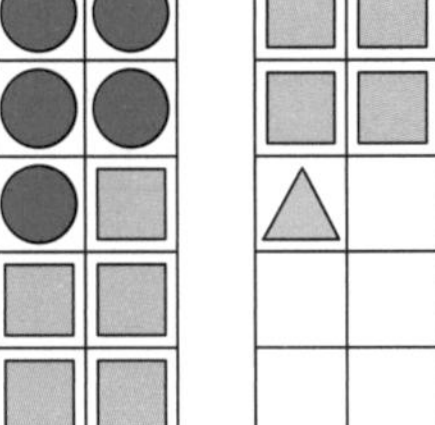

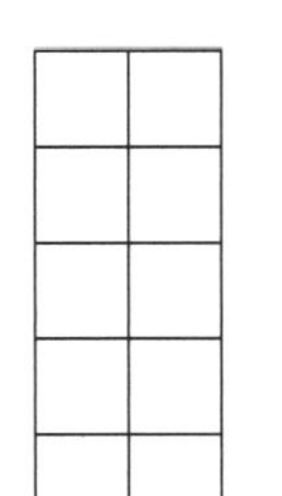

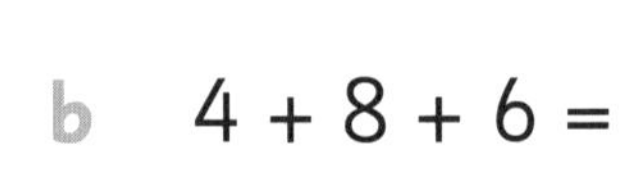

b 4 + 8 + 6 = ☐ + ☐ + ☐ = ☐☐

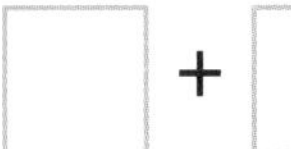

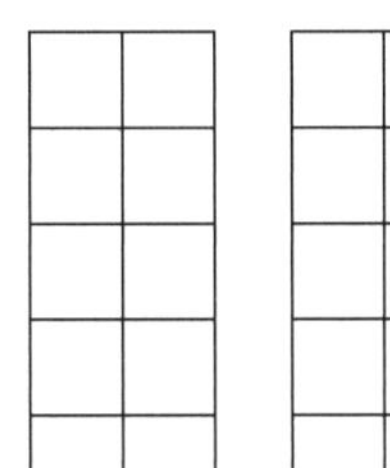

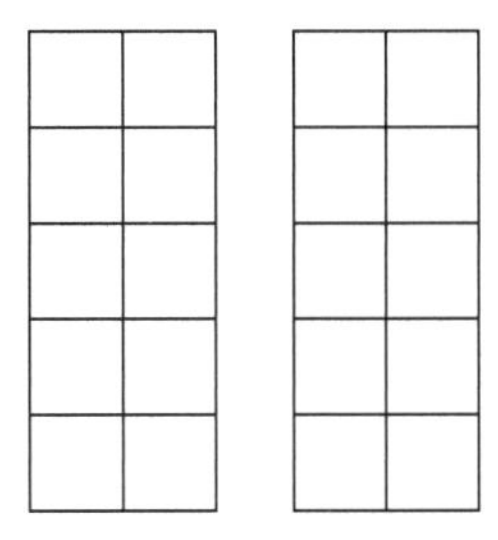

---

3 Rewrite and add.

3 + 6 + 7 = ☐ + ☐ + ☐ = ☐☐

Challenge

1 Show on the number line and solve.

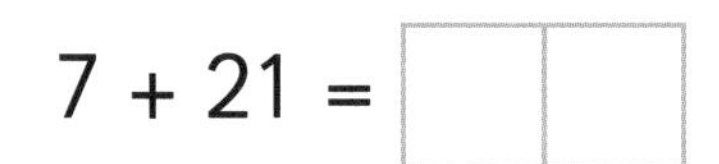

---

2 Solve on the number lines.

a Alex cycled 8 km to work and 31 km after work. How far did he go altogether?

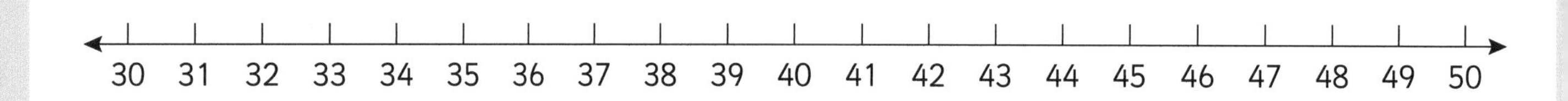

☐☐ + ☐ = ☐☐

b Liza's hens laid 25 eggs on Saturday and 33 eggs on Sunday. How many eggs did they lay over the weekend?

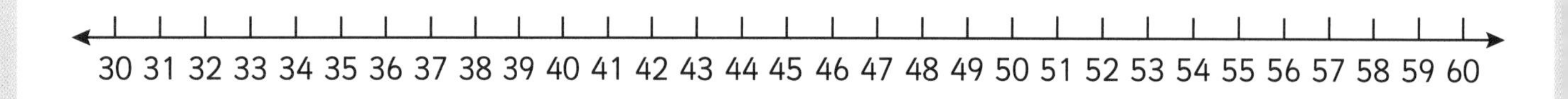

☐☐ + ☐☐ = ☐☐

---

3 Use the numbers to make three addition sums.

a 4, 6, 7, 17

☐ + ☐ + ☐ = ☐☐

☐ + ☐ + ☐ = ☐☐

☐ + ☐ + ☐ = ☐☐

b 42, 12, 17, 13

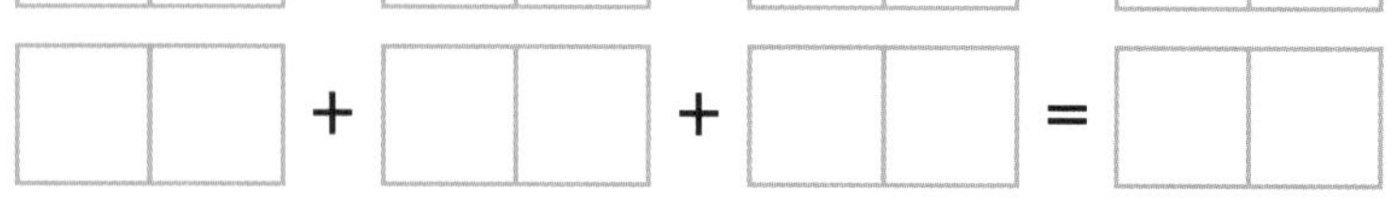

☐☐ + ☐☐ + ☐☐ = ☐☐

☐☐ + ☐☐ + ☐☐ = ☐☐

☐☐ + ☐☐ + ☐☐ = ☐☐

Mastery

1 This is a magic number square. Every row, column and diagonal adds up to the same number. You can only use each number once.

The magic number here is 15.

What number is missing from the centre square? ☐

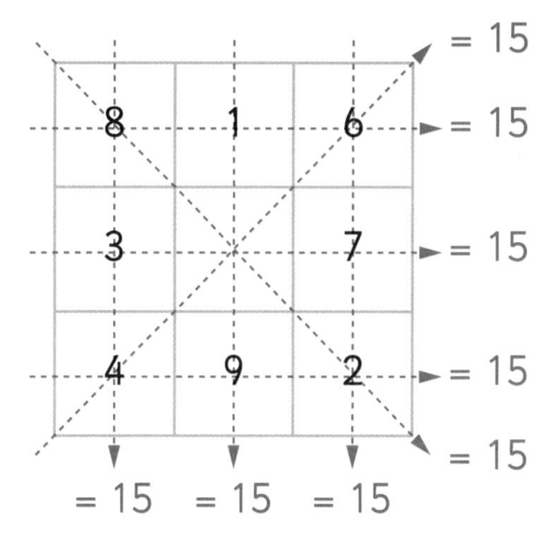

---

2 Fill in the gaps in these magic number squares, then write the magic number.

a

9	2	7
4		8
	10	3

The magic number is ☐.

b

8		4
6	10	14

The magic number is ☐.

c

		24
21		9
6		12

The magic number is ☐.

---

3 Try to solve these magic squares.

Hint: Look at the magic square at the top. How were the digits 1 to 9 used?

4		2

		8
		6

7		

---

4 There are also magic squares that are 4 squares wide by 4 squares high. Using the numbers from 1 to 16, the magic number is 34. Can you solve it? If it is too difficult, your teacher might give you permission to search online to solve the problem.

Subtracting in your head

Practice

1 Show getting to a 10 to solve the equations.

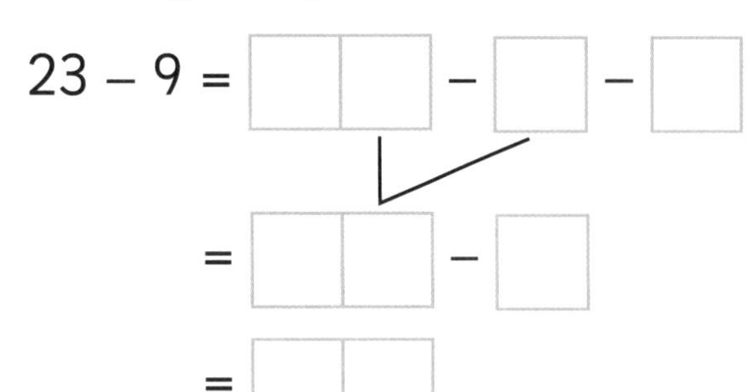

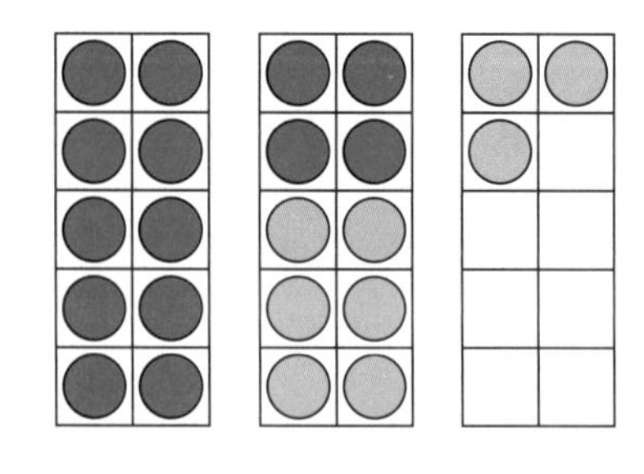

2 Solve by getting to a 10.

24 – 7 = ☐☐ – ☐ – ☐

= ☐☐ – ☐

= ☐☐

3 Count up to find the difference. Fill in the gaps.

25 – 18

18 + ☐ = ☐☐

20 + ☐ = ☐☐

So, 25 – 18 = ☐

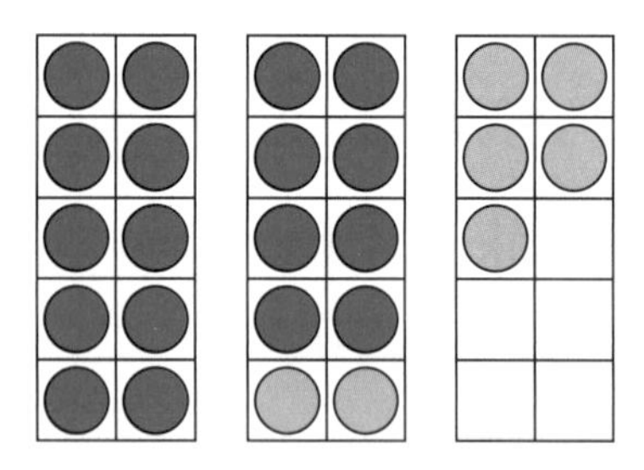

4 Count up to find the difference.

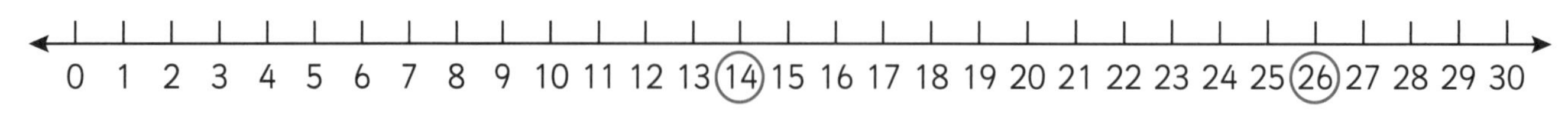

26 – 14 = ☐☐

Challenge

1 Use getting to a 10 to solve.

a 16 – 8 = ☐

b 17 – 9 = ☐

c 32 – 27 = ☐

d 42 – 33 = ☐

2 Count up to find the answers.

a 19 – 7 = ☐☐

b 29 – 16 = ☐☐

c 46 – 17 = ☐☐

d 48 – 23 = ☐☐

3 Choose your strategy to solve.

a Evie has $26. Tom has $15. How much more does Evie have?

b Jack has 16 books. Sam has 33 books. How many more does Sam have?

c Ava has 42 stickers. Jo has 24 stickers. How many more does Ava have?

Mastery

1 Write a subtraction story to match the picture.

2 Tess has 23 lollies. She used to have a lot more. Write a subtraction story and matching number sentence to show how she ended up with 23.

3 Billy is getting a new bike. His mum finds the same bike in two shops.

Which shop has the cheaper bike? ______

4 Fill the boxes to make subtraction sentences. Find two different solutions for each.

a

3 ☐ − 6 = 2 ☐

3 ☐ − 6 = 2 ☐

b

6 ☐ − 4 ☐ = ☐ 0

6 ☐ − 4 ☐ = ☐ 0

Oxford University Press

UNIT 1: TOPIC 6
Exploring subtraction

Practice

1 Use the number lines to find the answers.

a 23 – 9 = ☐☐

0 1 2 3 4 5 6 7 8 9 10 11 12 13 14 15 16 17 18 19 20 21 22 23 24 25 26 27 28 29 30

b 46 – 18 = ☐☐

25 26 27 28 29 30 31 32 33 34 35 36 37 38 39 40 41 42 43 44 45 46 47 48 49 50

2 Fill in the gaps to show that addition and subtraction are connected.

a

0 1 2 3 4 5 6 7 8 9 10 11 12 13 14 15 16 17 18 19 20 21 22 23 24 25 26 27 28 29 30

18 + ☐ = 27 ☐☐ – 9 = 18

9 + ☐☐ = ☐☐ 27 – ☐☐ = ☐

b

25 26 27 28 29 30 31 32 33 34 35 36 37 38 39 40 41 42 43 44 45 46 47 48 49 50

☐☐ + 8 = 44 44 – 8 = ☐☐

8 + ☐☐ = ☐☐ 44 – ☐☐ = ☐

3 Write a matching subtraction fact.

17 + 8 = 25 ☐☐ – ☐ = ☐☐

Challenge

1 Show on the number lines and solve.

a $24 - 7 = \square\square$

b $49 - 23 = \square\square$

c $34 - 5 - 4 = \square\square$

2 Write addition and subtraction facts to match the pictures.

a

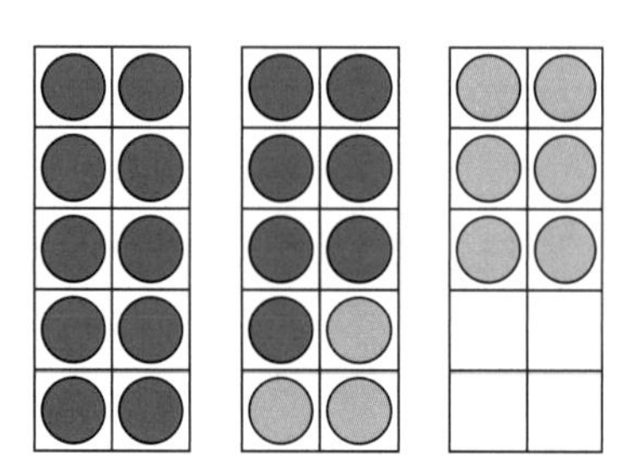

$\square + \square\square = \square\square$

$\square\square - \square = \square\square$

$\square\square + \square = \square\square$

$\square\square - \square\square = \square$

b

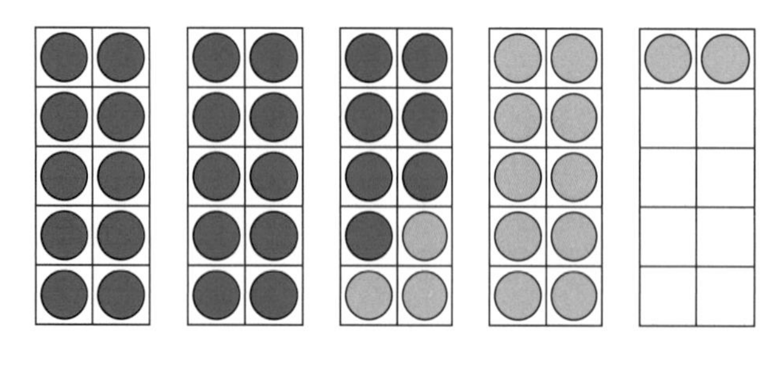

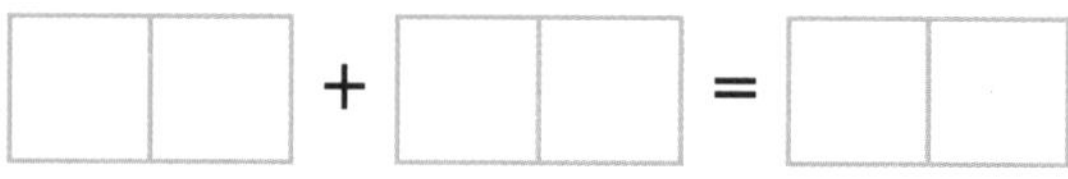

$\square\square + \square\square = \square\square$

$\square\square - \square\square = \square\square$

$\square\square + \square\square = \square\square$

$\square\square - \square\square = \square\square$

Mastery

1 If you put these number pairs on a number line, they would be the same distance apart.

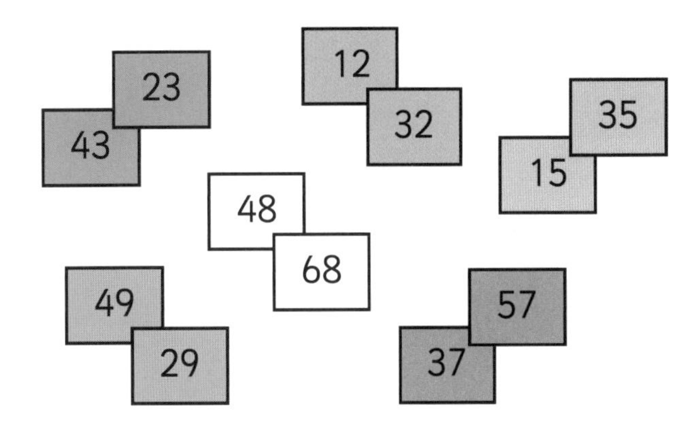

25 26 27 28 (29) 30 31 32 33 34 35 36 37 38 39 40 41 42 43 44 45 46 47 48 (49) 50

What is the difference between each pair of numbers? ☐

2 Fill the boxes with two-digit numbers so that there is a difference of 15 between each pair of numbers.

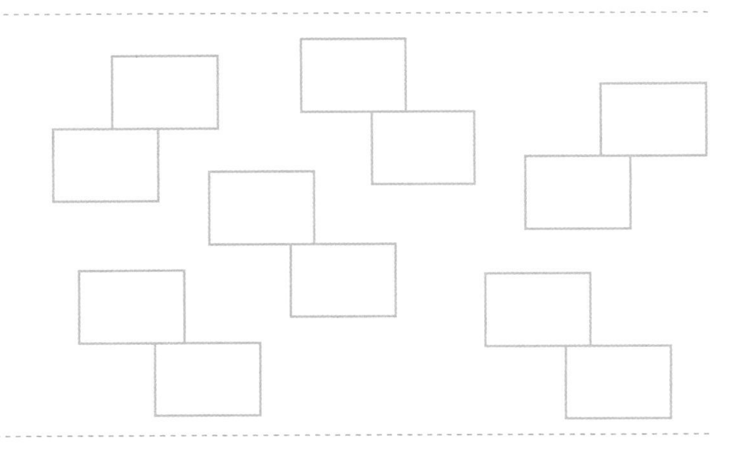

3 Henry has had a bad day. He has only 27 marbles in his bag. He knows he lost more than 15 marbles. Fill in the gaps to show how many he might have had to begin with.

He might have started with ☐ marbles.

He might have lost ☐ marbles.

He has 27 marbles left.

4 The dingo is 93 cm tall. The echidna is 28 cm tall. How much smaller is the echidna?

____

UNIT 1: TOPIC 7
Multiplying

Practice

1 Write the equation.

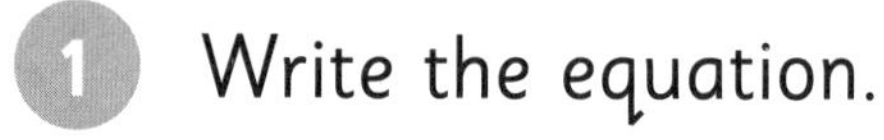

a ☐ × ☐ = ☐☐

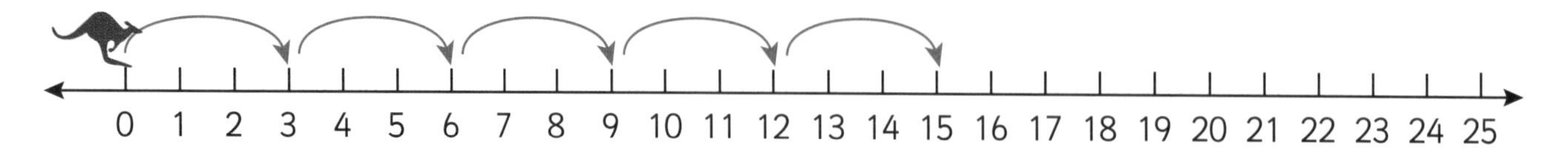

b ☐ × ☐☐ = ☐☐

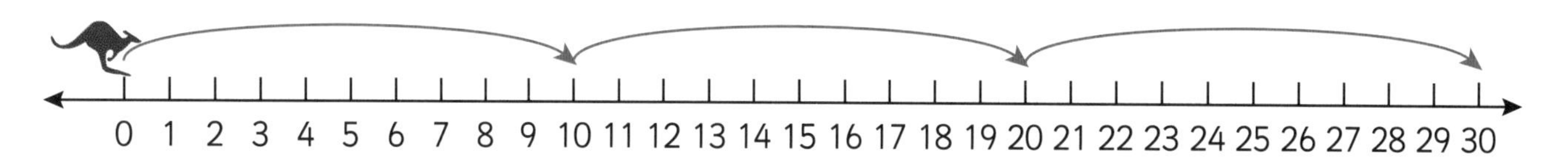

2 Show on the number line.

a 6 × 4 = ☐☐

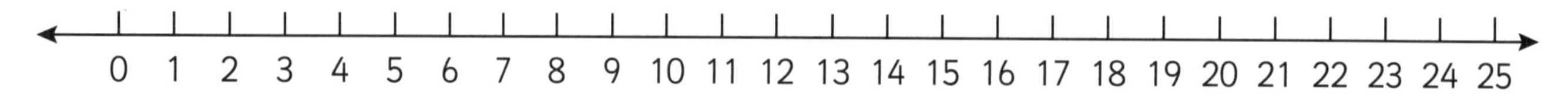

b 5 × 5 = ☐☐

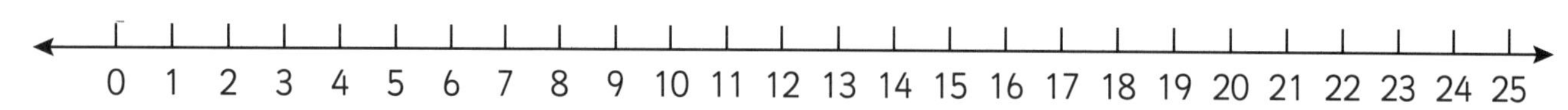

3 Use the number line to help you complete the 2s facts.

3 × 2 = ☐ 6 × 2 = ☐☐ 9 × 2 = ☐☐ 0 × 2 = ☐

0 1 2 3 4 5 6 7 8 9 10 11 12 13 14 15 16 17 18 19 20 21 22 23 24 25

4 Write an equation to match the array.

☐ × ☐ = ☐☐

Challenge

1 Draw arrays to show that 4 × 5 is the same as 5 × 4.

a 4 × 5 = 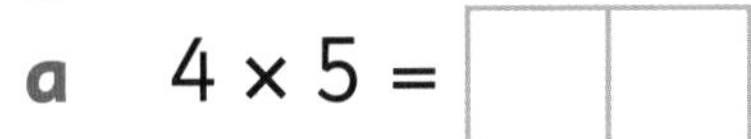

b 5 × 4 = 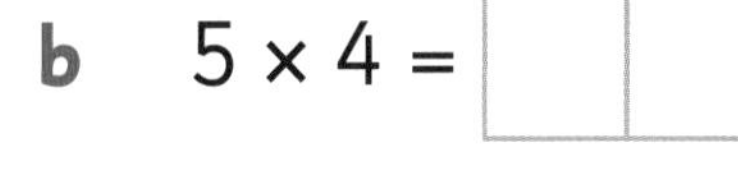

2 8 counters make an array of 8 rows of 1. Draw three more arrays of 8 counters.

☐ × ☐ = 8   ☐ × ☐ = 8   ☐ × ☐ = 8

3

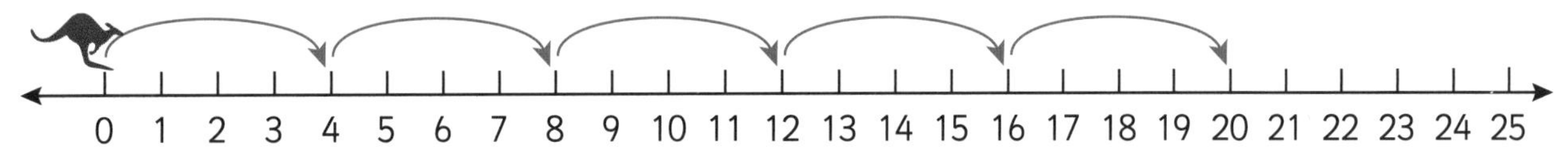

a If the kangaroo continues, will she land on 28? | Yes | No |

b How do you know?

Mastery

1 This array uses 24 counters.

$6 \times 4 = 24$

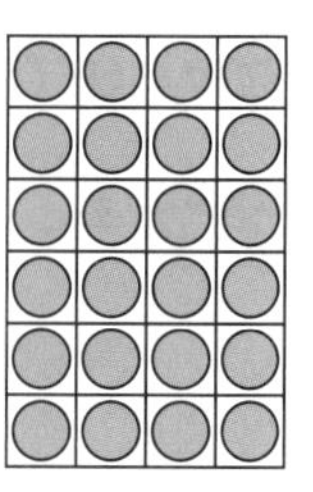

Use 24 counters and draw as many different arrays as you can. Write a number sentence for each.

2 Big Joey skips in 3s. He can only skip diagonally.

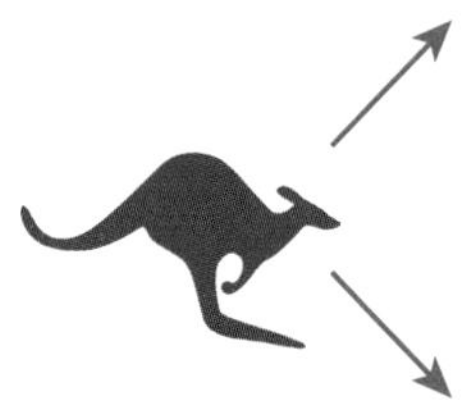

Little Joey skips in 2s. He can only skip forwards or sideways.

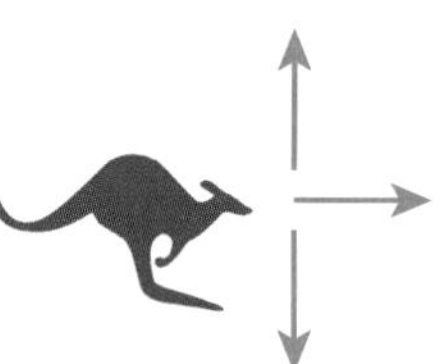

Draw a path for each joey across the paddock.

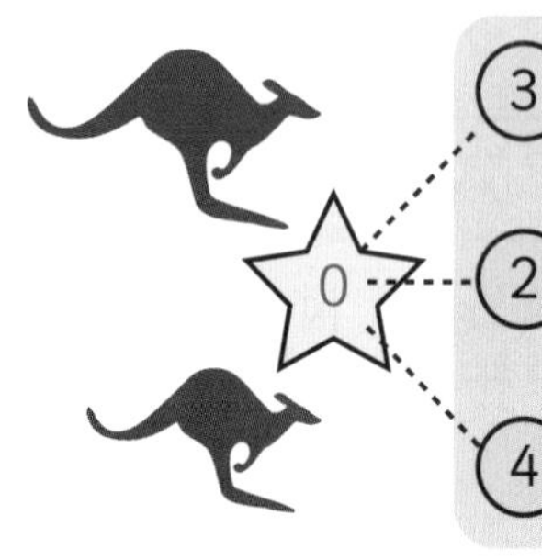

3	9	9	10	15	20	21
2	6	10	12	14	18	22
4	6	8	15	16	18	20

Dividing

Practice

1

a How many apples? ☐

b How many rows? ☐

c Fill in the blanks.

10 divided by 2 = ☐ 10 ÷ 2 = ☐

2 Share 10 flowers equally between 5 pots.

10 divided by 5 = ☐

10 ÷ 5 = ☐

3 Share the beads equally on the plates.

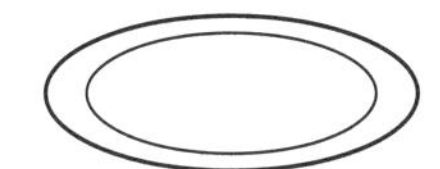
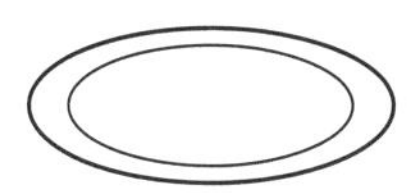
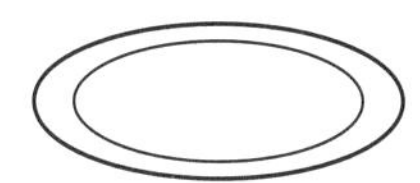

12 divided by 3 = ☐ 12 ÷ 3 = ☐

4 Draw counters to make the shares equal.

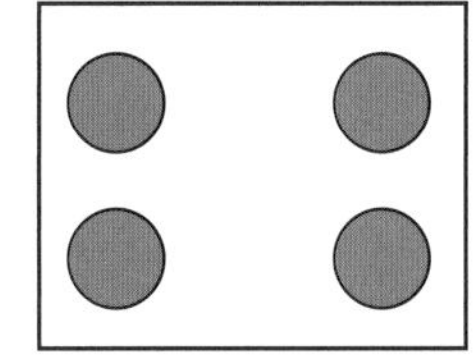
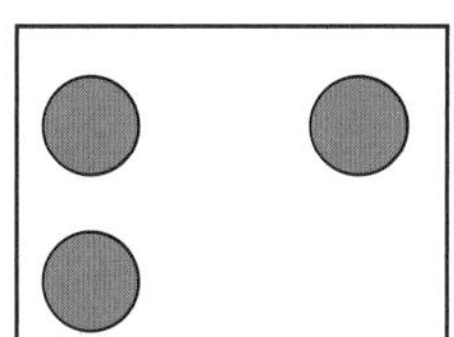
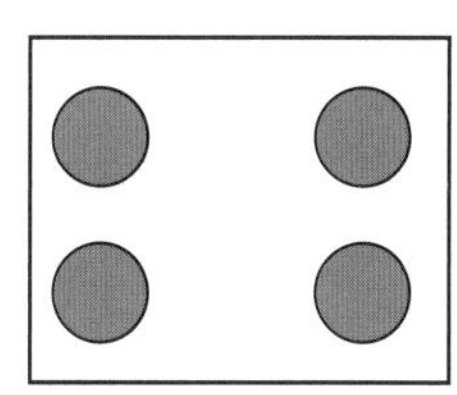
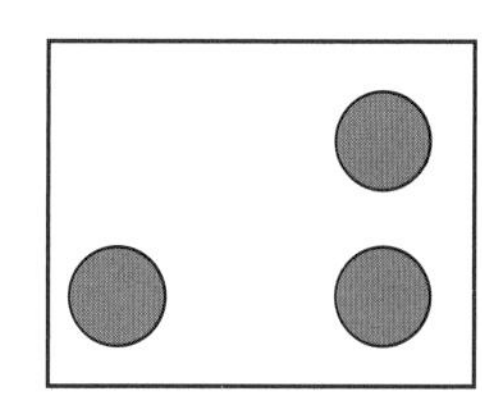

20 divided by ☐ = ☐ 20 ÷ ☐ = ☐

Challenge

1

a Share equally and write the equation.

b Share equally in different groups. Write the equation.

□ ÷ □ = □

2 12 stars could be shared like this:

$12 \div 2 = 6$

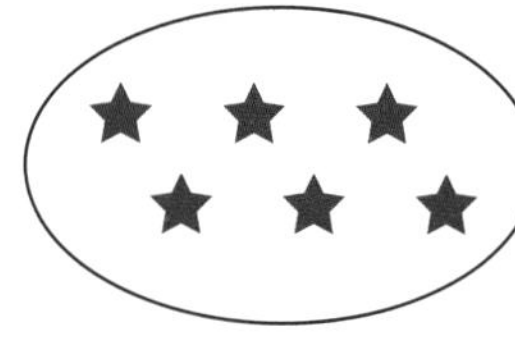

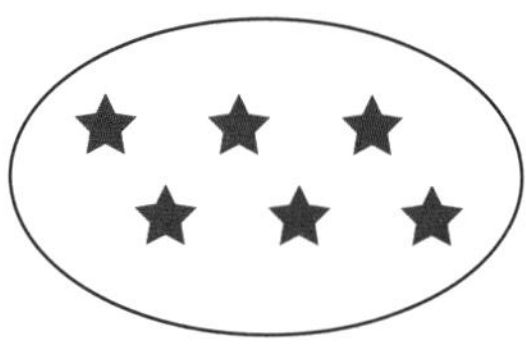

Draw and write three more ways to share 12 stars.

12 ÷ □ = □ 12 ÷ □ = □ 12 ÷ □ = □

3 Students are split into equal groups of 6.

How many students might be in the class?

Circle the correct answers.

12 16 18 20 24 26

Mastery

1 Mrs Khan has 3 children. They are given a box of 16 toy cars. Mrs Khan wants the children to share the toys.

a Explain why the toy cars cannot be shared equally between the 3 children.

____

____

b How do you think the problem can be solved?

____

____

2 40 lollies are put into 5 jars. Each jar has the same number of lollies.

a Draw a division story.

b How many lollies go in each jar?

c Write a division sentence for the story. ______________________

3 Try this Lucky Dip counters game.

- Players take turns.
- Take a handful of counters from a bag without looking.
- Find the number of counters.
- Score 1 point for each different array that can be made with the counters.

Practice

1 What fraction has been shaded?

a

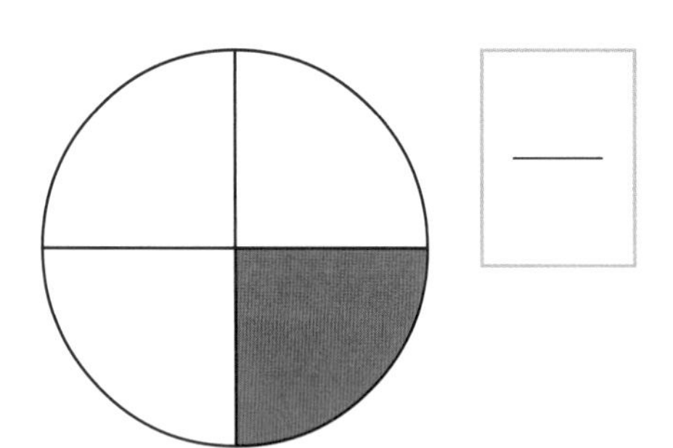

b

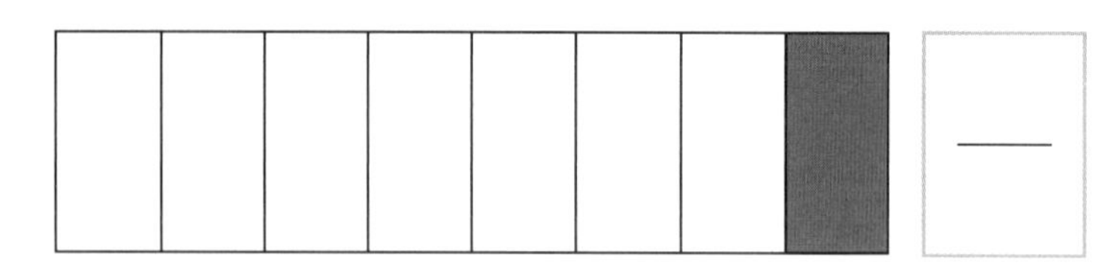

c

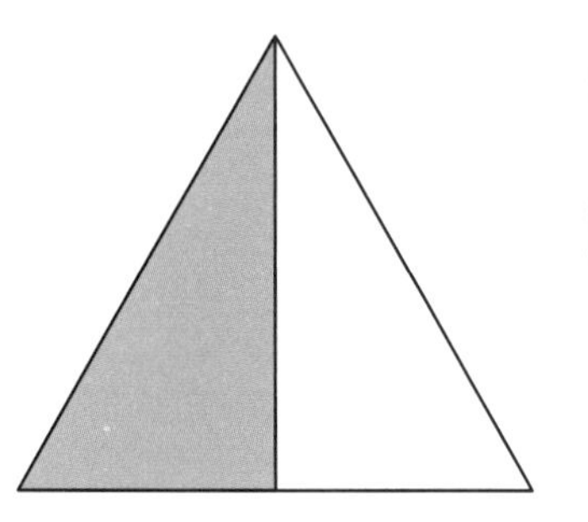

d

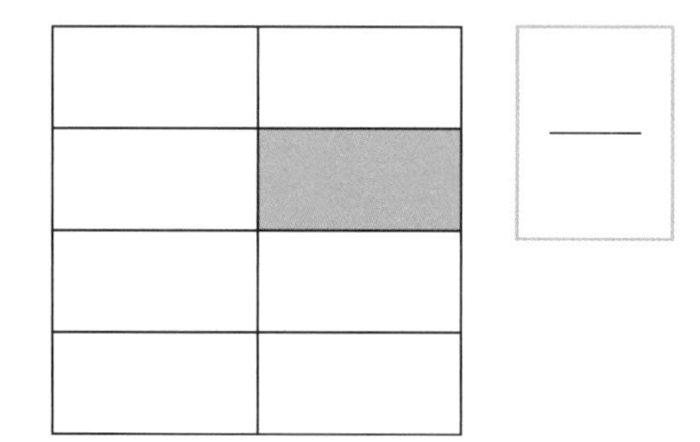

2

a Colour $\frac{1}{4}$ of the shape.

b Colour $\frac{1}{8}$ of the shape.

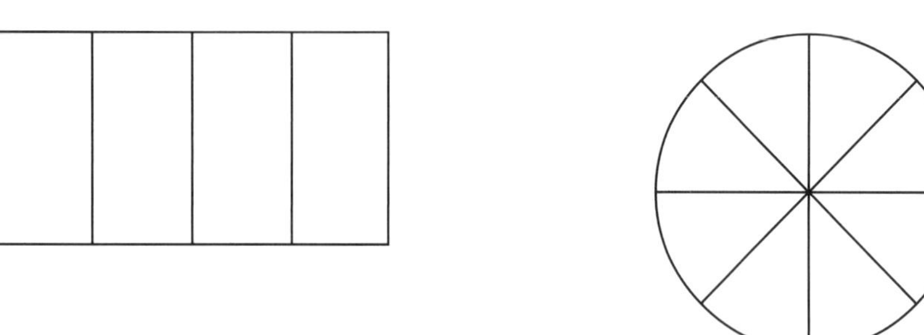

3 Draw a line from the fraction or whole number to its matching picture.

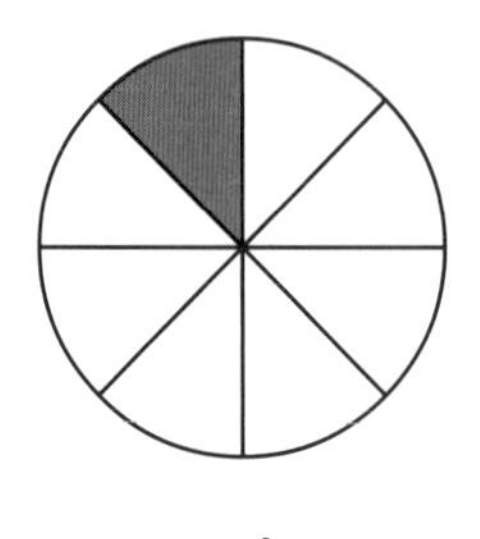

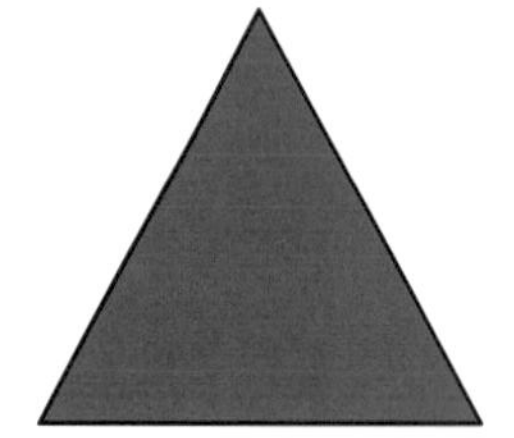

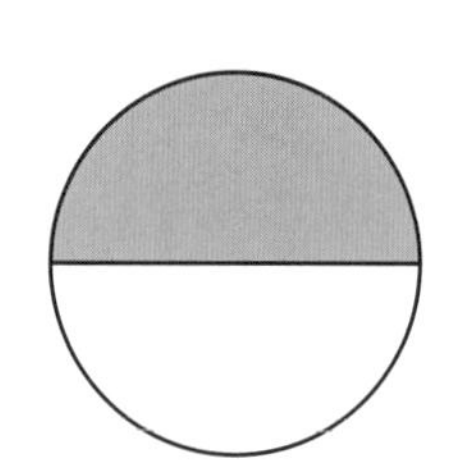

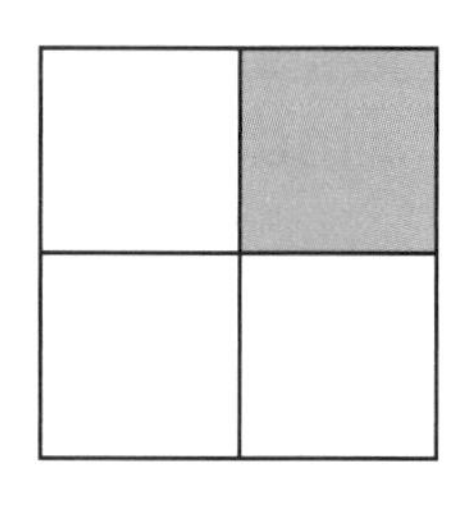

1 $\frac{1}{2}$ $\frac{1}{4}$ $\frac{1}{8}$

Challenge

a Shade $\frac{1}{2}$ red.

b Shade $\frac{1}{4}$ blue

c Shade $\frac{1}{8}$ green.

d What fraction is white?

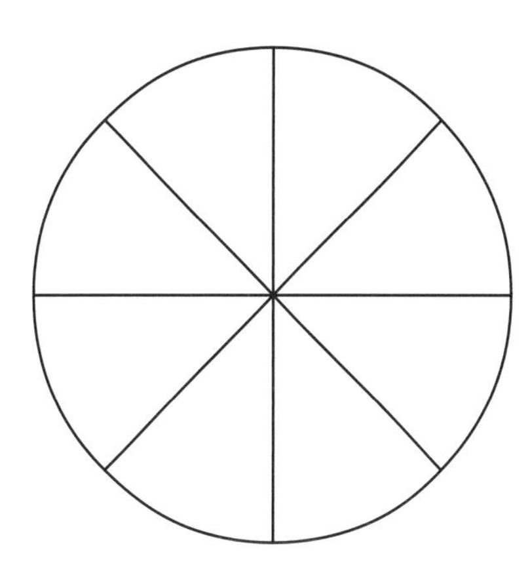

---

2 Divide into:

a quarters.

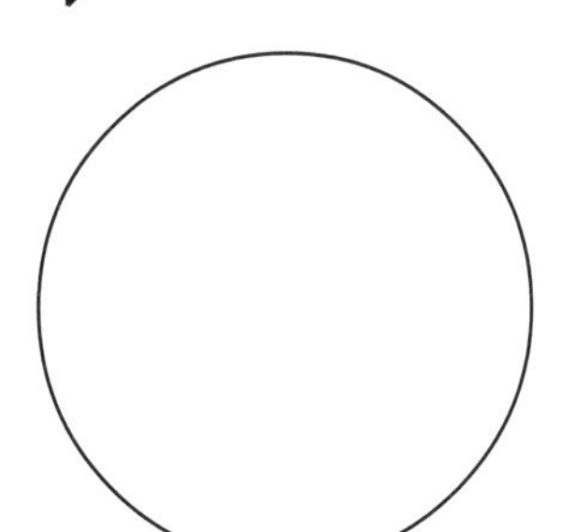

b eighths.

c halves.

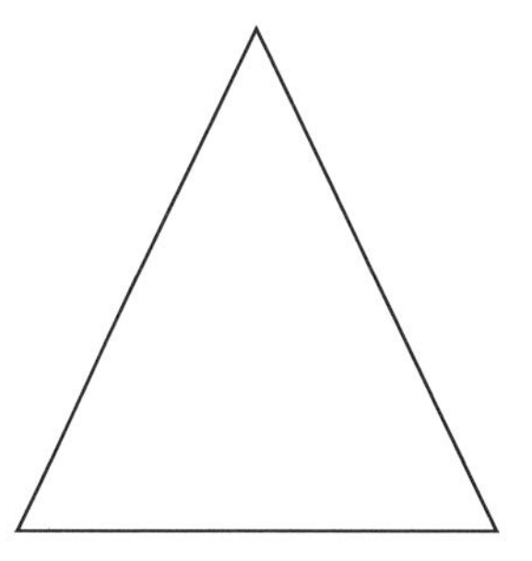

3

a Colour the shapes that are divided into quarters.

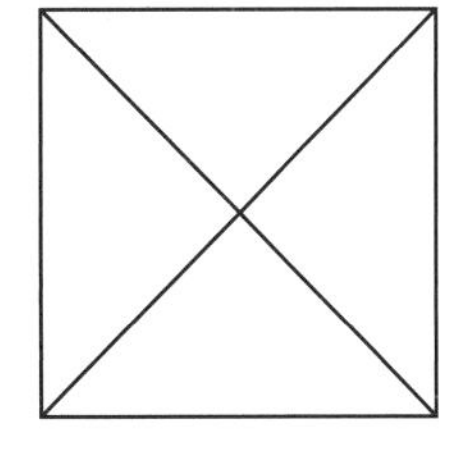
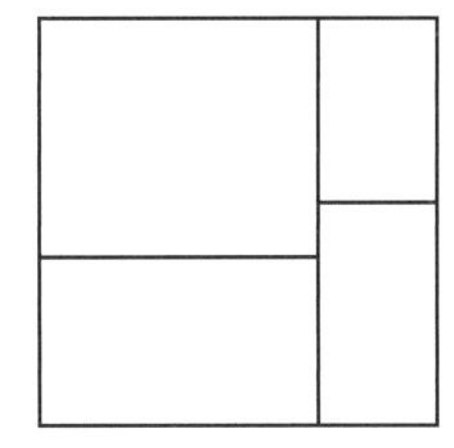
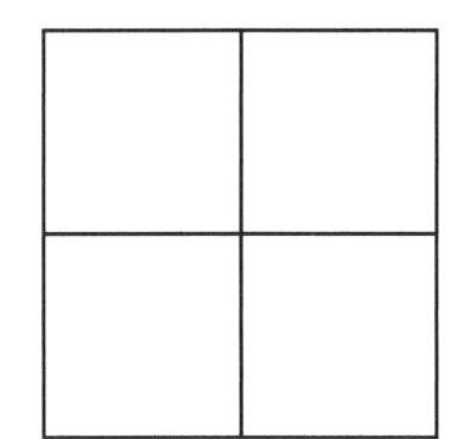

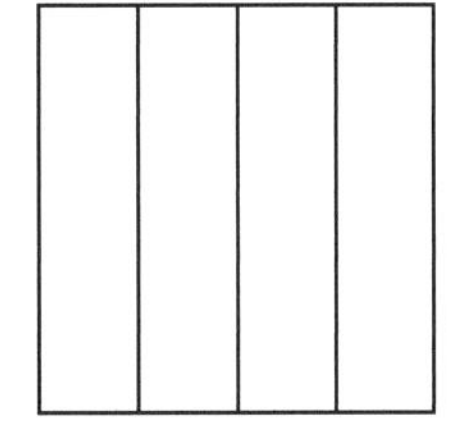
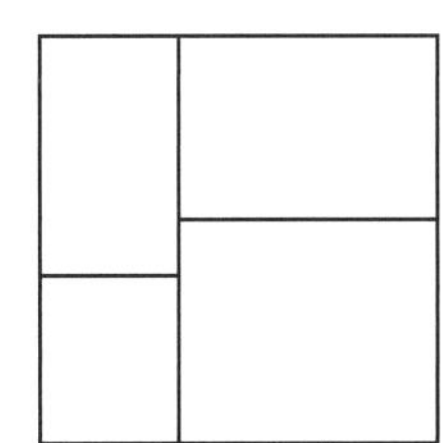
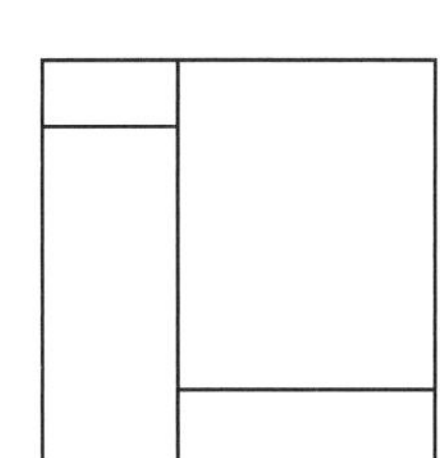

b How do you know that the other shapes are not divided into quarters?

Mastery

1 **a** Circle $\frac{1}{2}$ of this group in red.

b Circle $\frac{1}{4}$ of this group in blue.

c Circle $\frac{1}{6}$ of this group in green.

Which group is the smallest? ______________________

2 One quarter of the first shape has been shaded. Shade three quarters of the second shape.

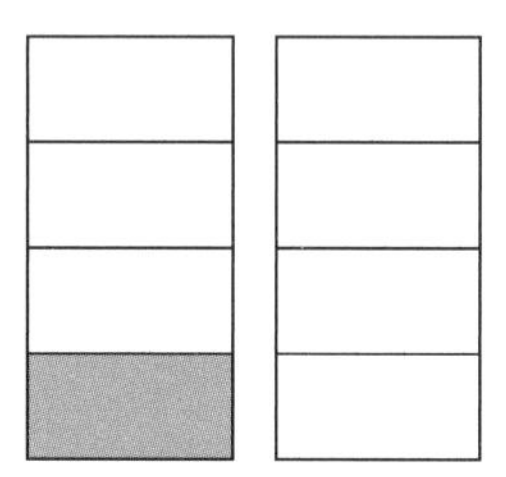

3 One eighth of the first shape has been shaded. Shade three eighths of the second shape.

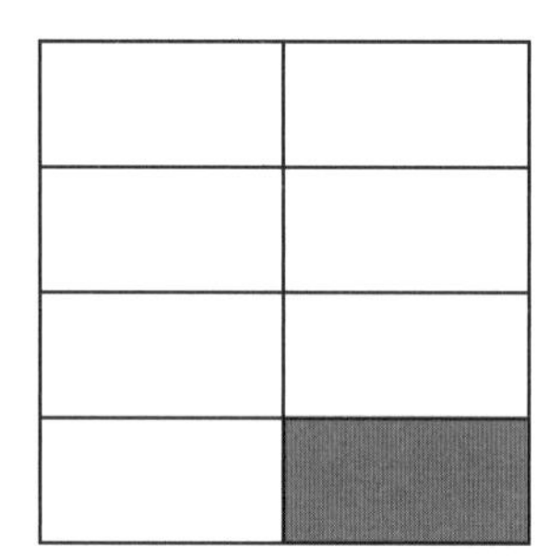

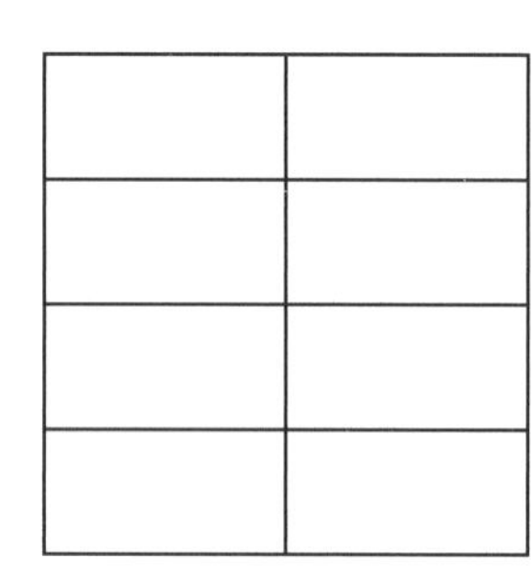

4 There is more than one way to divide a shape in half. Shade half of each shape. Colour them in an interesting way.

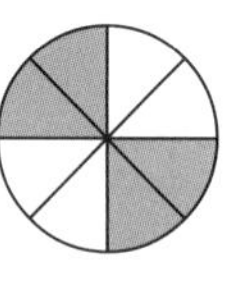

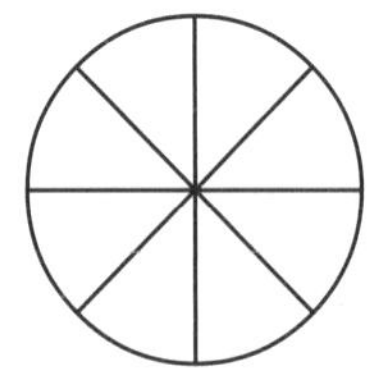

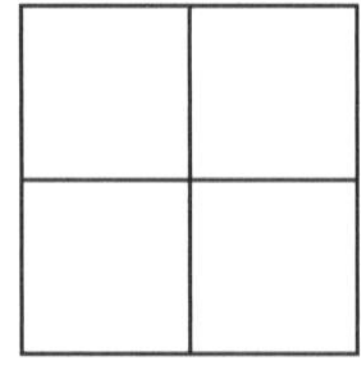

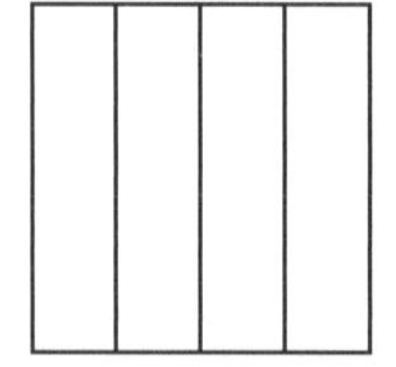

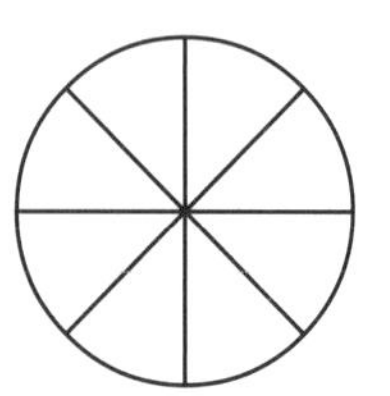

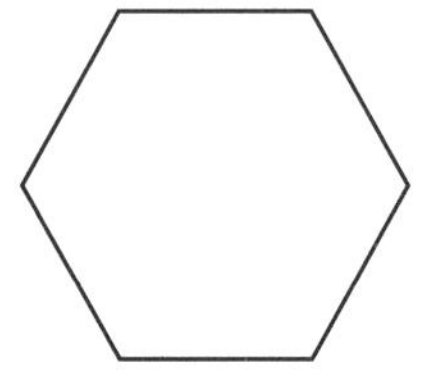

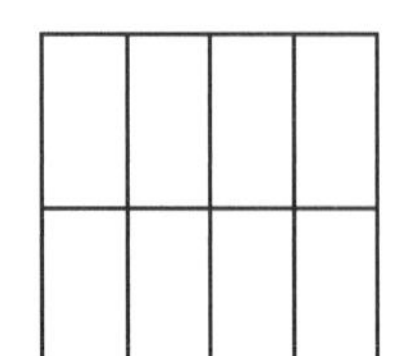

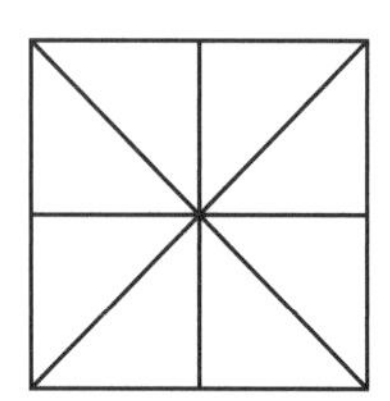

UNIT 2: TOPIC 2
Fractions of groups

Practice

1

a Colour $\frac{1}{4}$.

b Colour $\frac{1}{8}$.

2 What fraction is this?

a

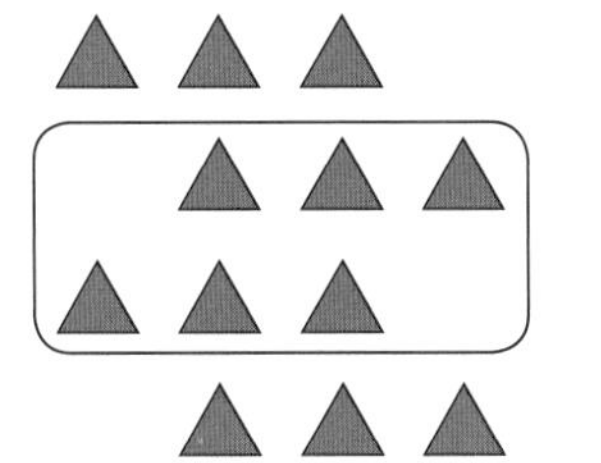

b

3

a Circle $\frac{1}{4}$.

$\frac{1}{4}$ of 8 is ☐.

b Circle $\frac{1}{4}$.

$\frac{1}{4}$ of ☐☐ is ☐.

4

a Colour $\frac{1}{8}$ red.

b Colour $\frac{1}{4}$ blue.

c Colour $\frac{1}{2}$ yellow.

d What fraction are white?

Challenge

a Draw hats on $\frac{1}{2}$.

b Make $\frac{1}{2}$ smiley and $\frac{1}{4}$ sad.

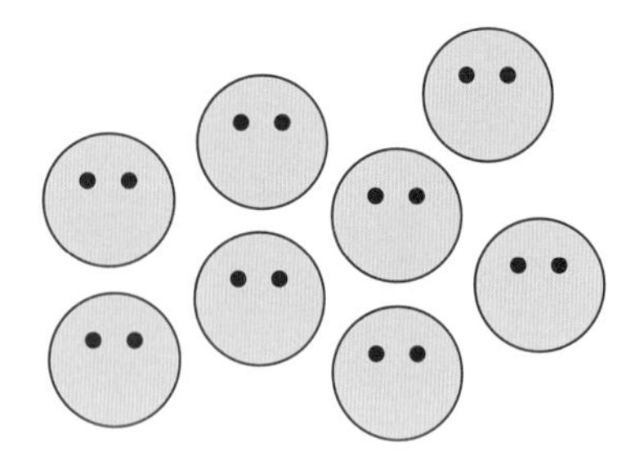

2 Circle which is bigger.

a $\frac{1}{4}$ of 16

OR

$\frac{1}{2}$ of 10

b $\frac{1}{4}$ of 12

OR

$\frac{1}{8}$ of 16

3 What fraction of the group of 24:

a have pink faces?

b have smiley faces?

c have yellow, smiley faces?

d are not sad or happy?

1 Ava likes jelly beans. Would she choose a quarter of the jar of 40 or a half of the jar of 16? Explain why.

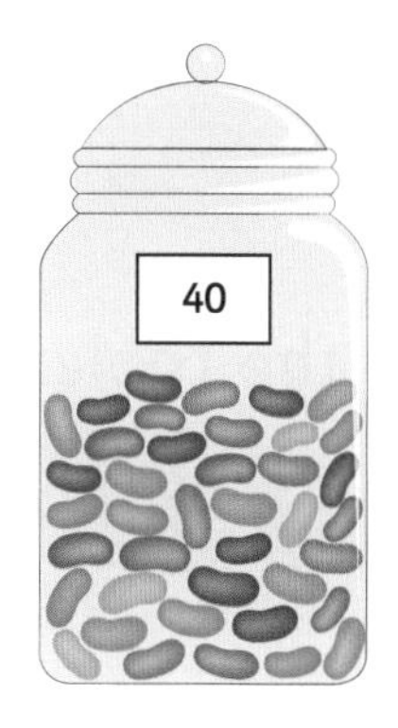

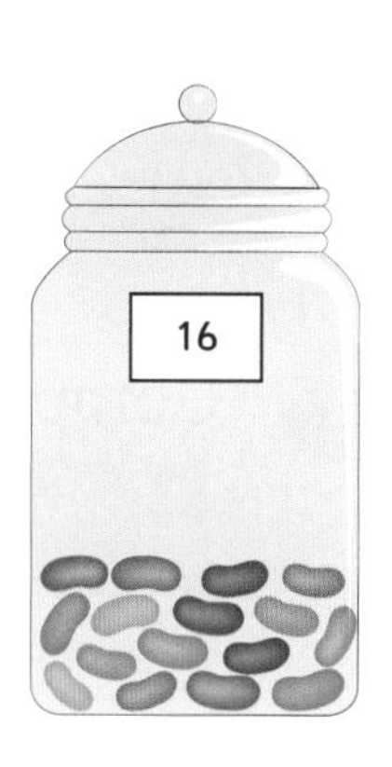

2 Write fraction stories about the faces. One has been done for you.

One half of the pink faces are happy.

3

a Divide the group using eighths, quarters and halves. Decorate the hats to show how you have split the group.

b Write some fraction stories about the group.

Notes and coins

Practice

1 Complete the table.

How many of these ...	... do you need to make this?	Draw the answer	Write the answer
5	20		
10	50		
50	1 DOLLAR		

2 Circle notes that equal:

a $35

b $65

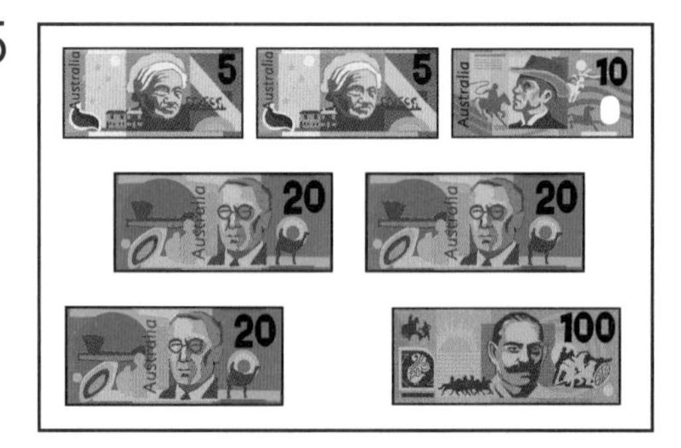

3 How much is this?

a

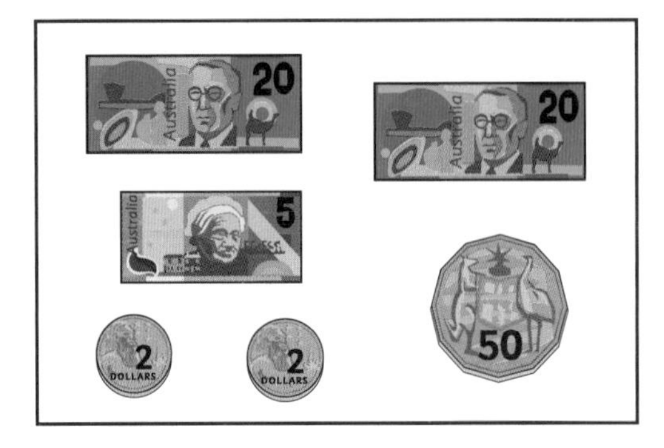

b

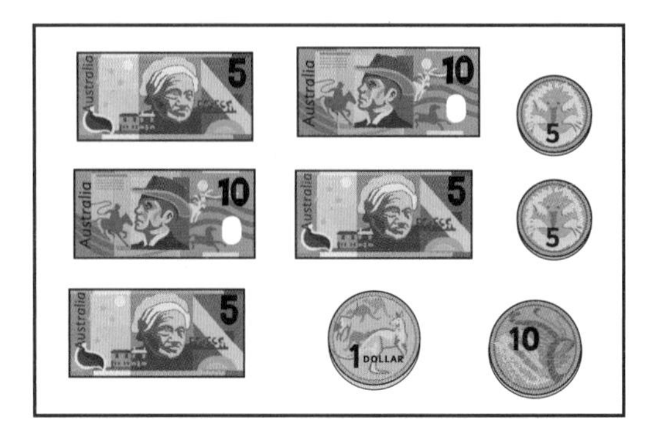

Challenge

1 Draw coins to make these amounts:

a 3 coins that make 15c

b 4 coins that make 50c

c 4 notes that make $25

d 1 note and 3 coins that make $6.25

2 Draw and count:

a one of each type of coin

b one of each type of note

3 Draw ways that coins can make 30c.

Mastery

1. Ava was given a $100 note for her birthday. She changed it for other notes. Draw and write three ways to make $100 in notes.

2. Ava emptied her piggy bank. She had twenty $1 coins. Draw and write three other ways that she could have $20 in notes and coins.

3. Ava bought a game for $2.50. She gave 2 coins and got 3 coins in change.

   a Draw the coins that she might have paid with.

   b Draw the coins that she might have got in change.

UNIT 3: TOPIC 2
Counting money

Practice

1 Count how much.

a

b

2 Make 60c with:

a 20c coins

b 5c coins

3 Make $200 with:

a $50 notes

b $20 notes

Challenge

1 Draw the smallest number of coins you could use to make:

a $1.80

Number of coins:

b $8.35

Number of coins:

2

a How much?

b How much would you have left if you spent:

i $20?

ii $35?

iii $53?

3

a How much?

b How much would you have left if you spent:

i $20?

ii $69?

iii $65?

Mastery

Shops often show prices such as $7.99, but we should round the price to $8.

1 Round these prices up or down.

a 

b 

c 

d $5.01

e 

f 

---

2 Round the prices of each item.

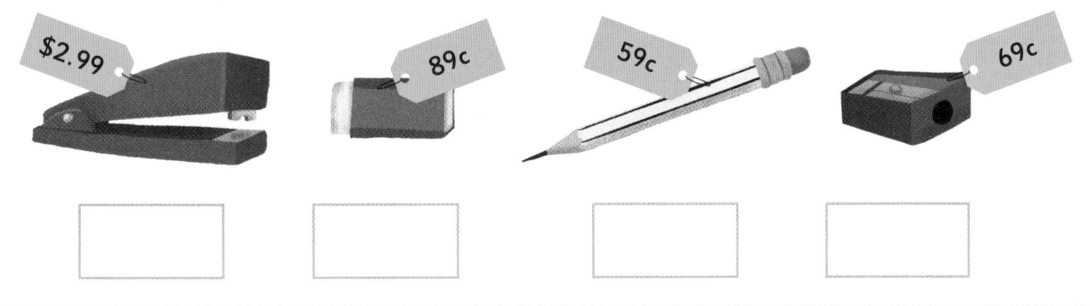

---

3 Look at the items in question 2. Show the change amount if you bought:

a an eraser with a $1 coin.

b a sharpener with a $2 coin.

c a pencil with a $5 note.

d a stapler with a $20 note.

e a stapler and a pencil with a $10 note.

f a pencil and a sharpener with a $2 coin.

UNIT 4: TOPIC 1
Describing patterns

Practice

1

a Circle the final digits in the pattern.

10	20	30							

b The pattern is counting by: ☐

c Complete the pattern.

2

a Circle the final digits in the pattern.

40	36	32	28	24					

b The pattern is counting by: ☐

c Complete the pattern.

3 Find the missing numbers.

a

3		9	12			21			

b

1		5	7		11				

c

51	46	41				21			

d

101		81		61					

Challenge

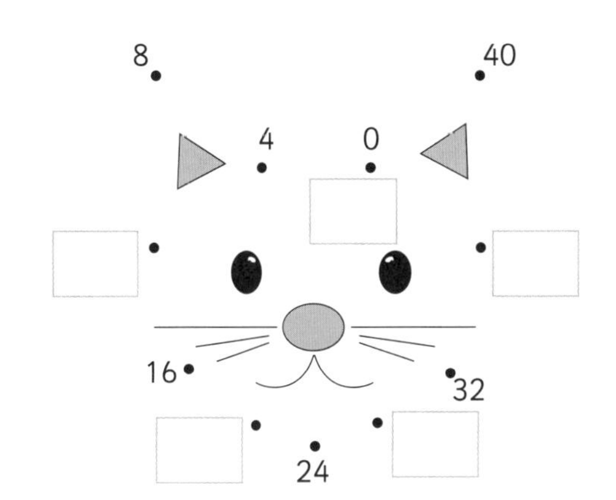

a What is the counting pattern?

b Fill in the missing numbers.

c Join the dots by following the pattern.

---

2 Half of 32 is 16.

a Circle 16 in red in the chart. Then circle half of 16 in red.

b Continue the halving pattern by circling the answer to 1a in a different colour and circling its half in the same colour.

c Use the same process to complete the pattern.

0	1	2	3	4	5	6	7	8	9	10
	11	12	13	14	15	16	17	18	19	20
	21	22	23	24	25	26	27	28	29	30
	31	32	33	34	35	36	37	38	39	40

---

3 Use the chart in question 2 to help you answer the questions.

a What is half of 16? ______________

b What is the answer if you halve 16, then halve it again? ______________

c What is double 8? ______________

d What is the answer if you double 8, then double it again? ______________

Mastery

1

a Choose a number to count by and shade the hundred chart.

b What number did you choose? ________

c Choose a different number to count by and shade with a different colour.

d What was your second number? ________

1	2	3	4	5	6	7	8	9	10
11	12	13	14	15	16	17	18	19	20
21	22	23	24	25	26	27	28	29	30
31	32	33	34	35	36	37	38	39	40
41	42	43	44	45	46	47	48	49	50
51	52	53	54	55	56	57	58	59	60
61	62	63	64	65	66	67	68	69	70
71	72	73	74	75	76	77	78	79	80
81	82	83	84	85	86	87	88	89	90
91	92	93	94	95	96	97	98	99	100

e Choose a third number to count by and circle the numbers in the counting pattern.

f What was your third number? ________

g What numbers are in each of the first two patterns?

____

h What numbers are in each of the second and third patterns?

____

i What numbers are in all three patterns?

____

2 Make a simple dot-to-dot picture using a different number pattern than the one on page 39. Use a separate piece of paper.

Word problems

Practice

1 Write a number sentence for the word problem.

a Mia ate 9 cherries. Then she ate another 5. How many did she eat altogether?

Number sentence:

---

b There are 8 cupcakes and 3 candles. How many more candles are needed?

Number sentence:

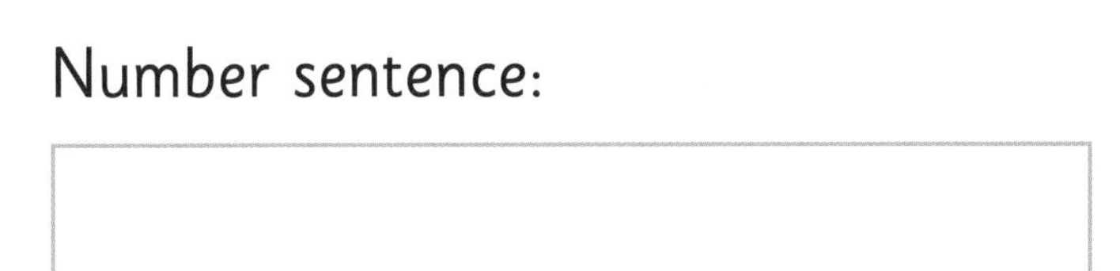

---

2 Draw the problem, then write a number sentence to solve it.

a Kim blew up 14 balloons. He burst 6 of them. How many balloons were left?

Picture

Number sentence:

b Tilly has 7 marbles. She gets another 8 marbles. How many does she have altogether?

Picture

Number sentence:

Challenge

1. Write a word problem and number sentence to match each picture.

Word problem

Number sentence

2.

Word problem

Number sentence

3. Draw a diagram to match the word problem. One monster has three heads. How many heads do three of these monsters have?

Write the answer as a number. ____________

Mastery

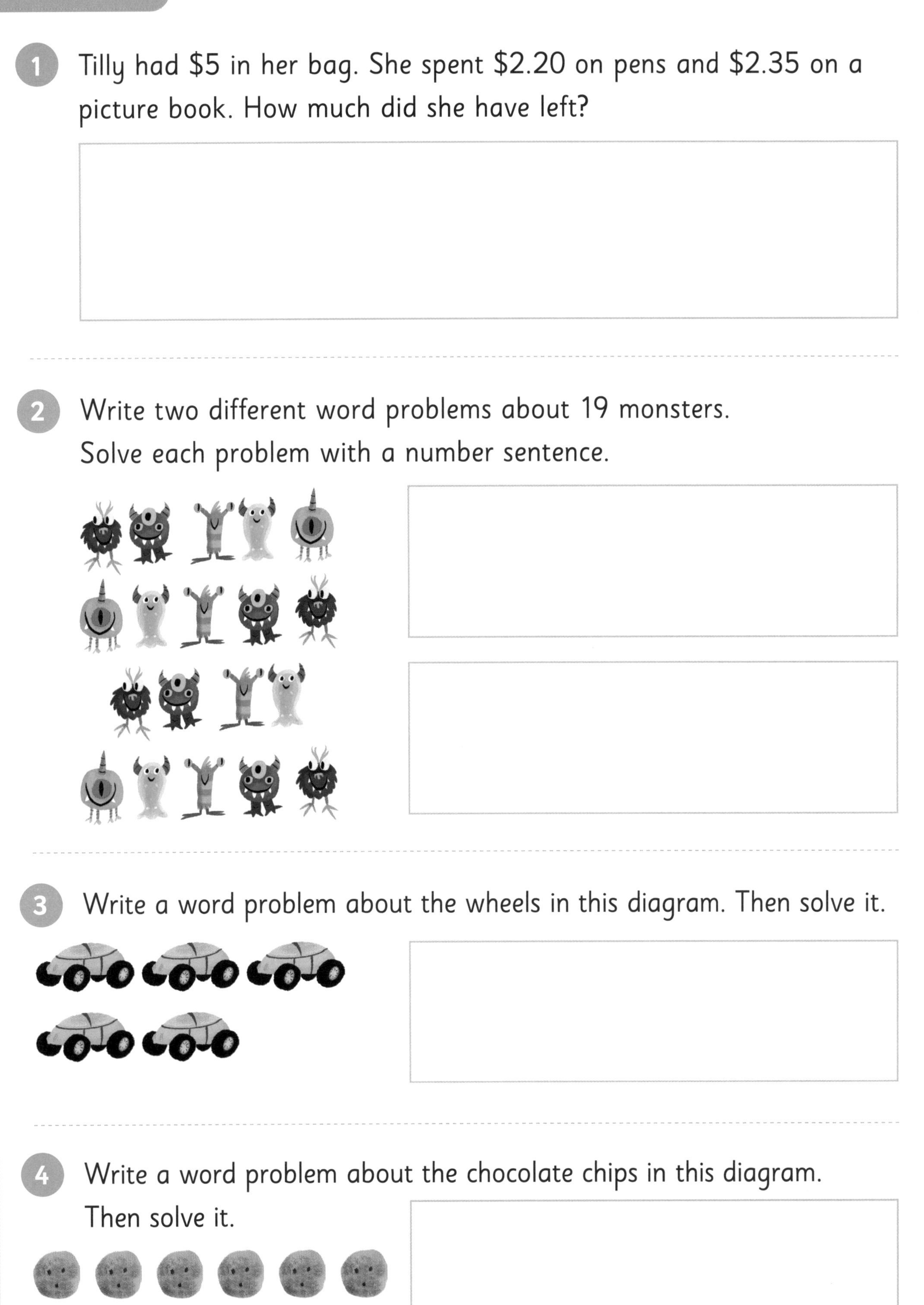

1. Tilly had $5 in her bag. She spent $2.20 on pens and $2.35 on a picture book. How much did she have left?

2. Write two different word problems about 19 monsters. Solve each problem with a number sentence.

3. Write a word problem about the wheels in this diagram. Then solve it.

4. Write a word problem about the chocolate chips in this diagram. Then solve it.

Practice

a Choose a unit of measure to complete the table.

hand span

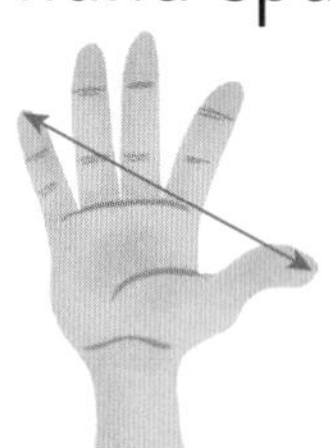

finger length

finger width

Length to find	Unit of length	Estimate	Actual length
my leg			
a texta			
a sharpener			
my chair height			

b Why did you choose that unit? ______________________

____

2 Choose one of the units pictured to find the area of the items below.

a Estimate the area of each item.

b Find the area of each item.

c Complete the table.

sticky note

Maths book

Area to find	Unit of area	Estimate	Actual area
a table			
a reading book			
a chair seat			

Challenge

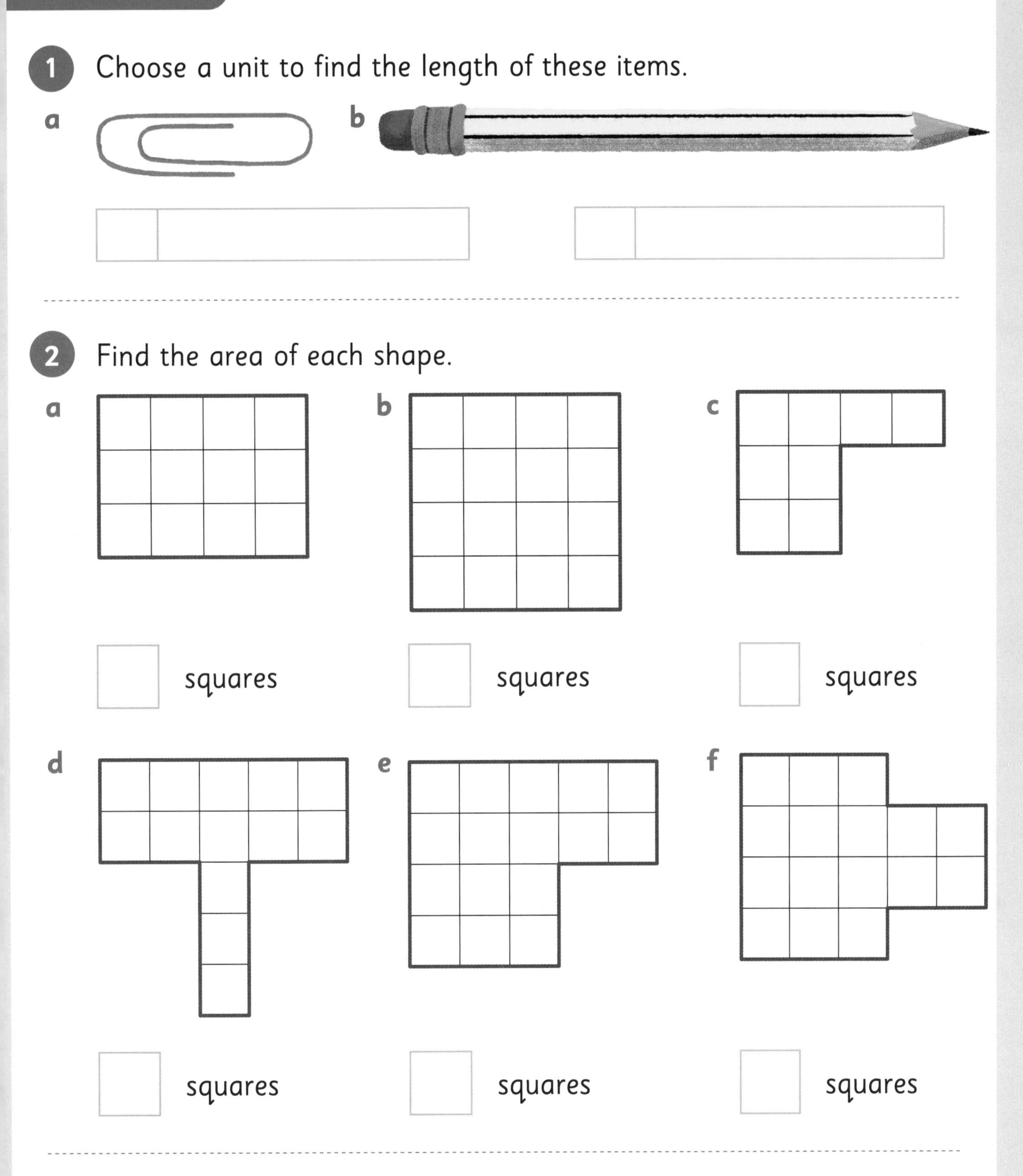

1 Choose a unit to find the length of these items.

a

b

2 Find the area of each shape.

a ☐ squares

b ☐ squares

c ☐ squares

d ☐ squares

e ☐ squares

f ☐ squares

3 Circle the shapes in question 2 that have the same area.

Mastery

To measure longer lengths we can use the length of a stride.

1 Choose a unit to measure the lengths and complete the table.

Length to find	Unit of length	Estimate	Actual length
from my table to the teacher's table			
a bookshelf			

2 Find the area of each rectangle.

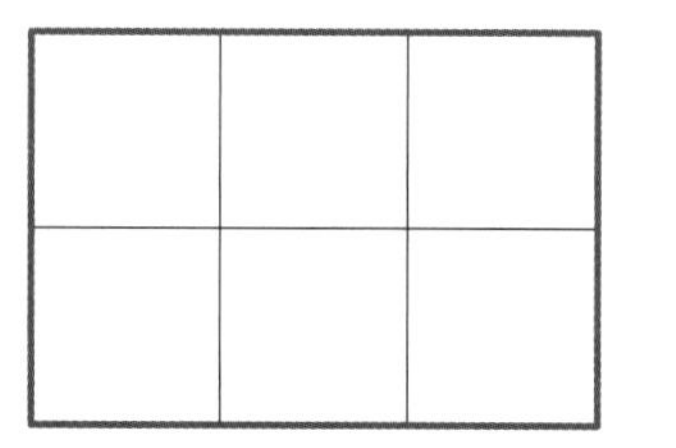

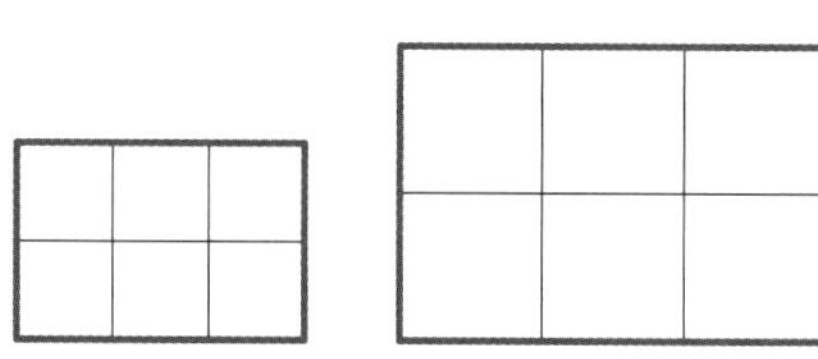

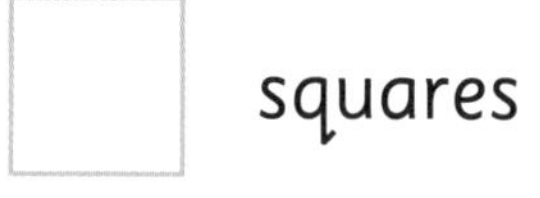

☐ squares ☐ squares ☐ squares

3

a Find the area of each robot.

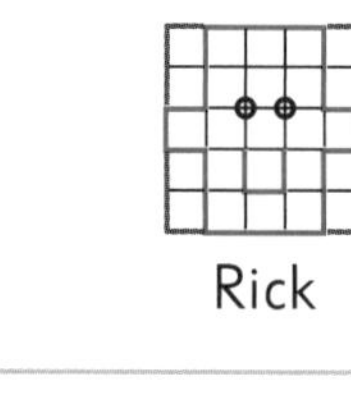

Rick

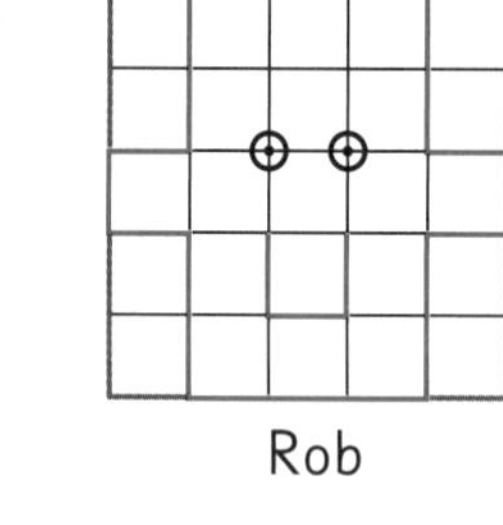

Rob

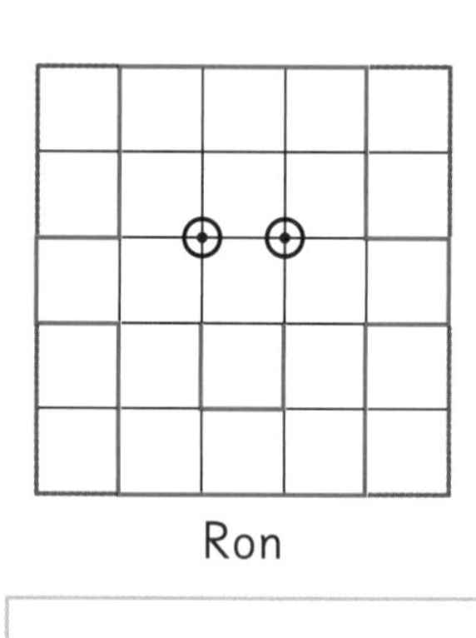

Ron

b Write the names in order, from the smallest robot to the largest.

____

c The **number** of squares for each robot is the same. But the **area** of each robot is not the same. Explain why.

____

Metres and centimetres

Practice

1 Estimate if these are less than a metre, about a metre or more than a metre.

	Your hand	A Year 2 person	A table	The classroom	A pencil
Less than, about or more than 1 m?					

2

a Estimate the lengths and write in the table.

Item	My estimate in metres	Actual length in metres
the whiteboard		
the width of the doorway		
the width of the room		
the height of a window		

b Use a metre ruler to find the actual lengths. Complete the table.

3 Use the centimetre grid to find the lengths of:

a the paperclip

☐ centimetres

b the eraser.

☐ centimetres

Challenge

1 Use a 30 cm ruler to find the length of these pencils.

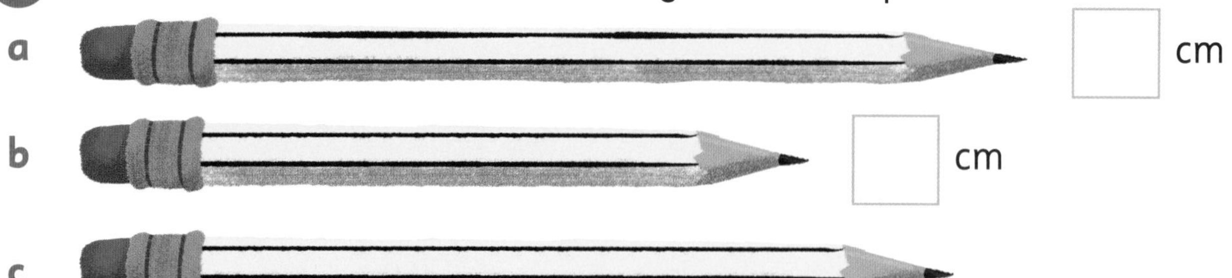

a ☐ cm

b ☐ cm

c ______

2 Which unit of length would you use for the following?

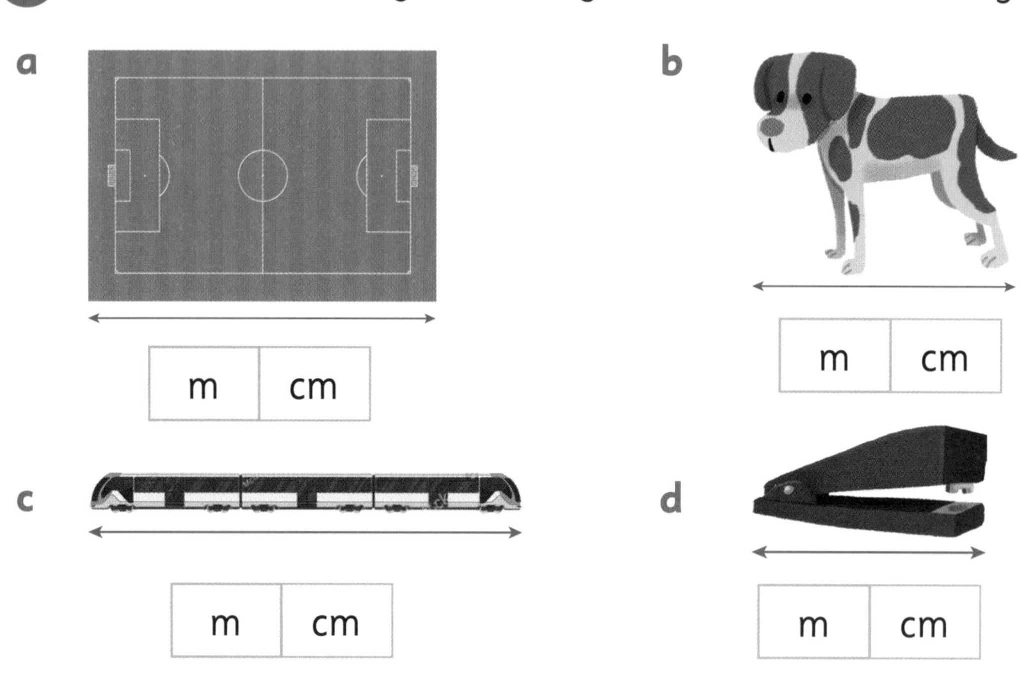

a | m | cm |

b | m | cm |

c | m | cm |

d | m | cm |

3 Circle the best estimate for the lengths of the objects in question 2.

a	the dog	3 cm	3 m	30 cm	30 m
b	the train	40 cm	40 m	1 cm	1 m
c	the soccer field	1 cm	1 m	100 cm	100 m
d	the stapler	15 cm	15 m	50 cm	50 m

Mastery

Objects can also be measured to the half centimetre.

This eraser is 4 and a half centimetres long.

1 Write the lengths of these pencils to the nearest half centimetre.

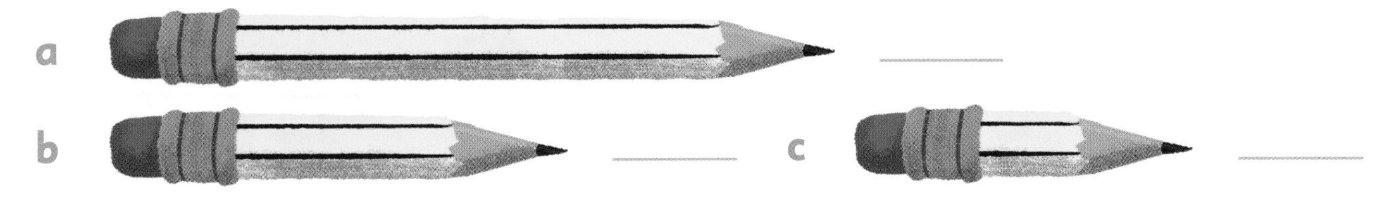

a ______

b ______ c ______

2 Write the heights of these items to the nearest half metre.

a Adult: ______

b Cow: ______

c Child: ______

d Dog: ______

e Kangaroo: ______

3 Use a ruler and pencil to measure and draw the lines. Start from each dot.

a 8 cm long

•

b 10 cm long

•

c 8 and a half cm long

•

d 10 and a half cm long

•

4 What is something in your classroom that measures closest to 1 and a half metres?

______

Volume and capacity

Practice

1 What is the volume of each object?

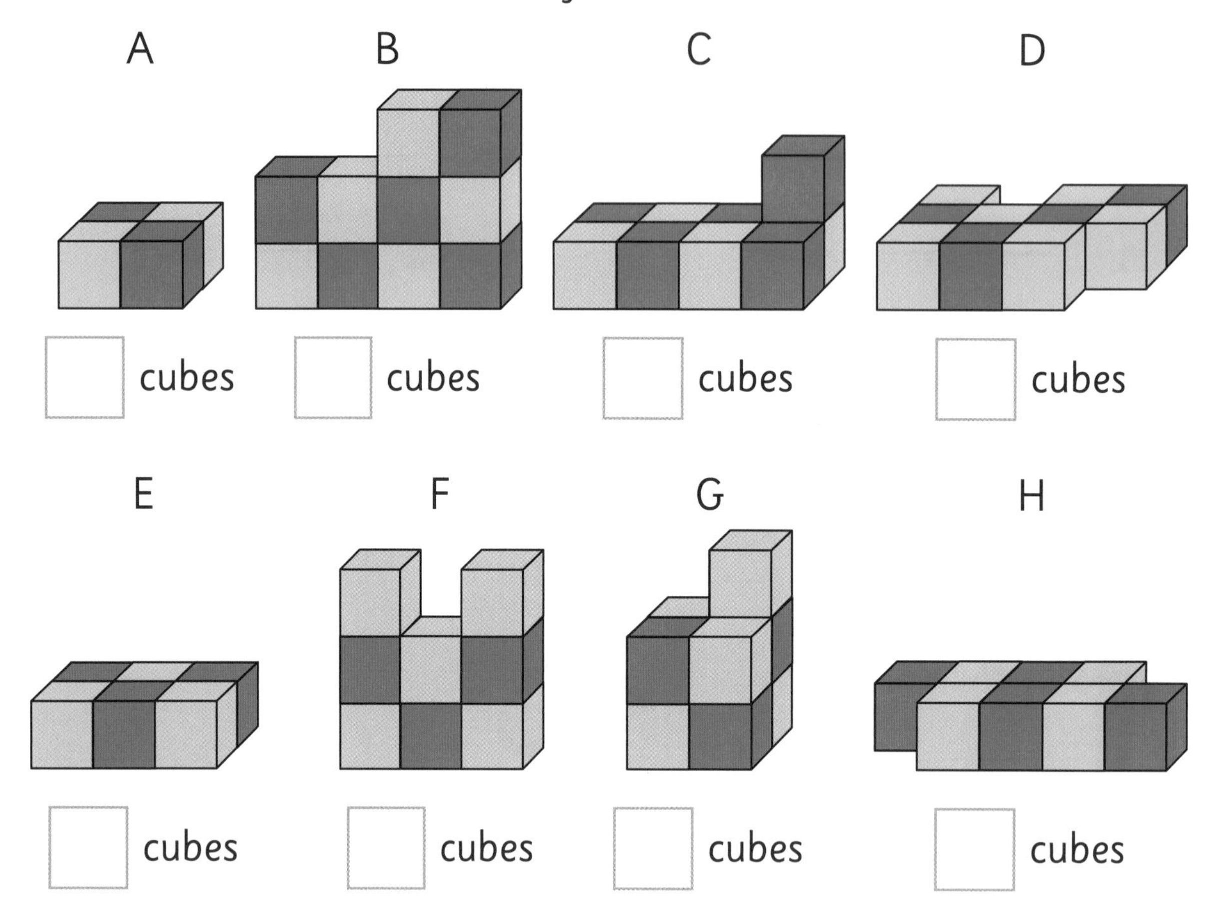

2 Write the letter of each container from smallest to largest capacity.

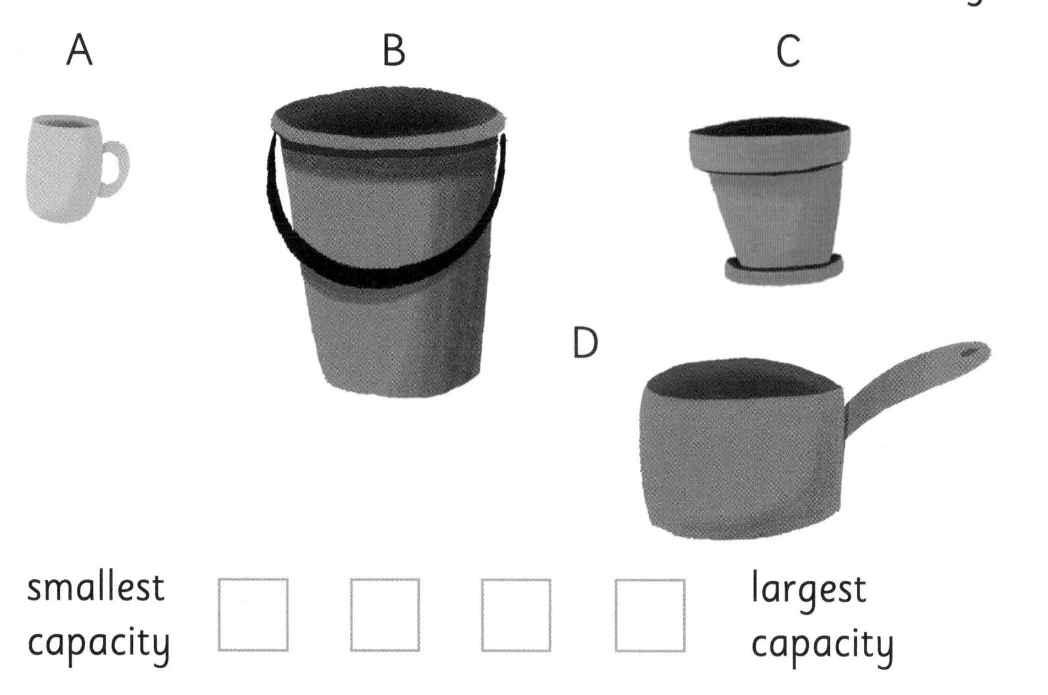

Challenge

1

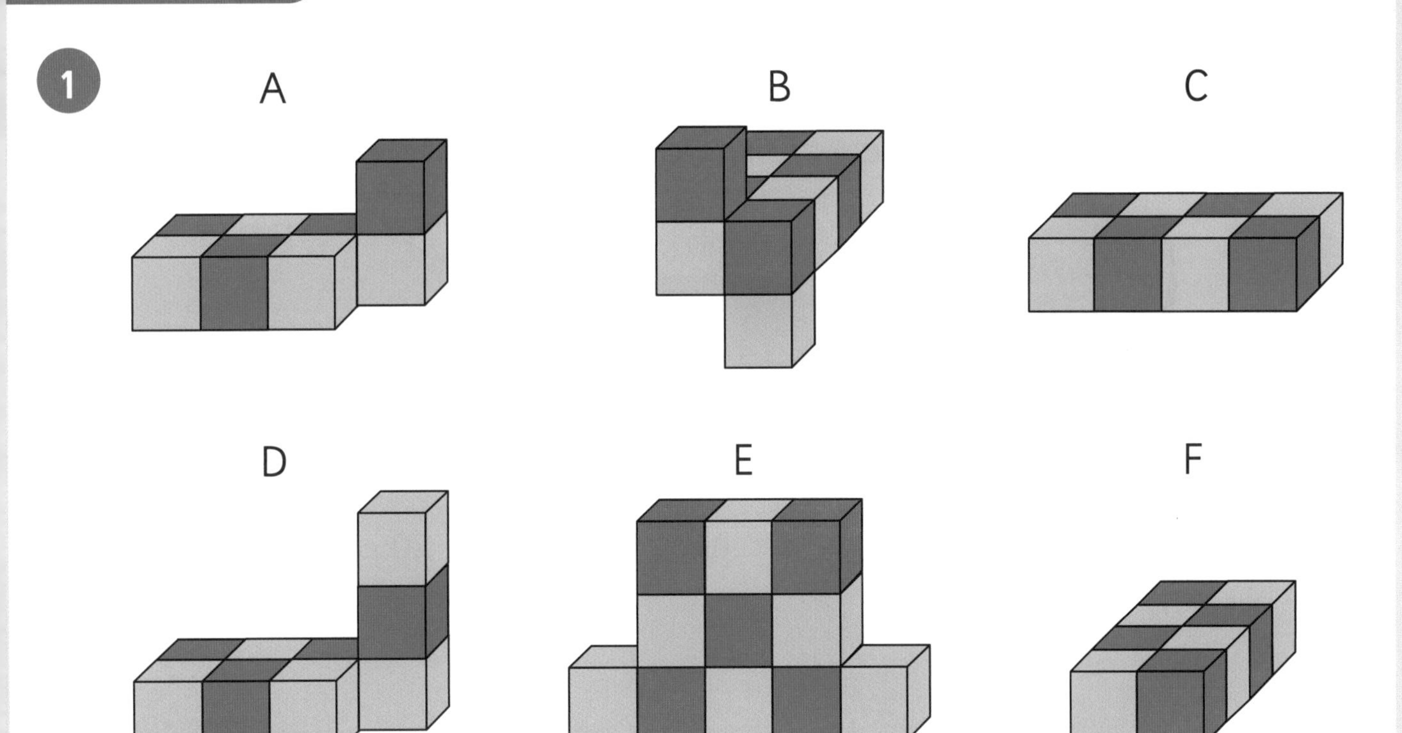

a Draw a square around the object with the biggest volume.

b Circle the objects with a volume of 8 cubes.

c Tick the object that has a volume that is 1 cube less than object B.

d What is the volume of object D? ____________

---

2 Number the containers 1 to 6 from smallest to largest capacity.

A B

C

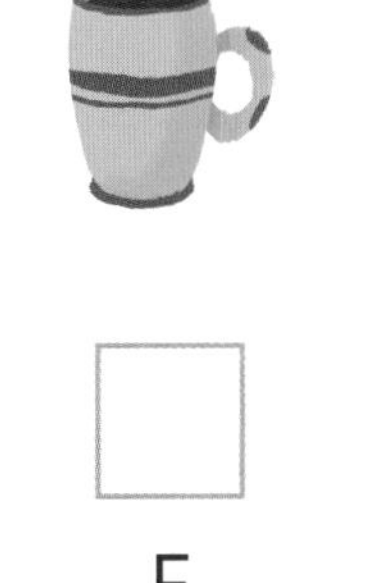

☐ ☐ ☐

D

E

F

☐ ☐ ☐

Mastery

1

a Circle the better estimate for the capacity of each container.

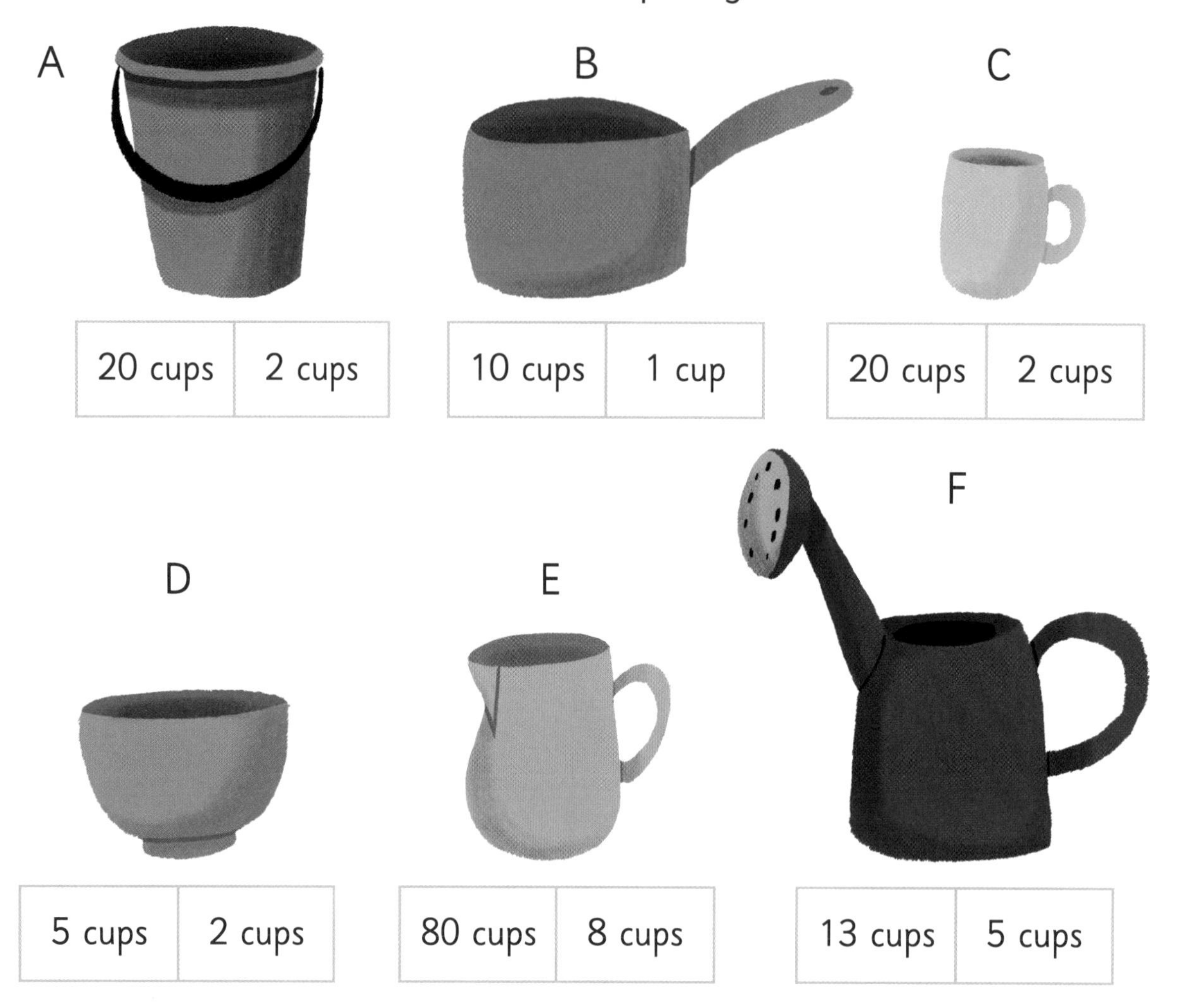

b Choose a container similar to one shown. Use a cup to find its real capacity. Write what you found out.

____

2

a What is the volume of this cube model?

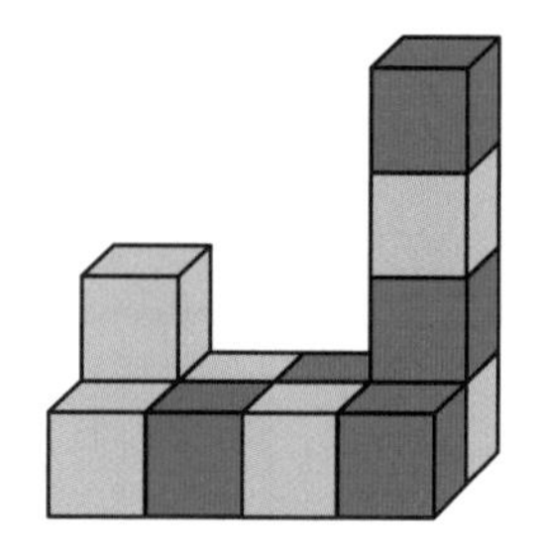

____

b Build three different models using the same number of blocks. Draw your models on a separate piece of paper.

Practice

1 Is the mouse **heavier** than, **lighter** than, or the **same** mass as the other objects? Circle the correct answer.

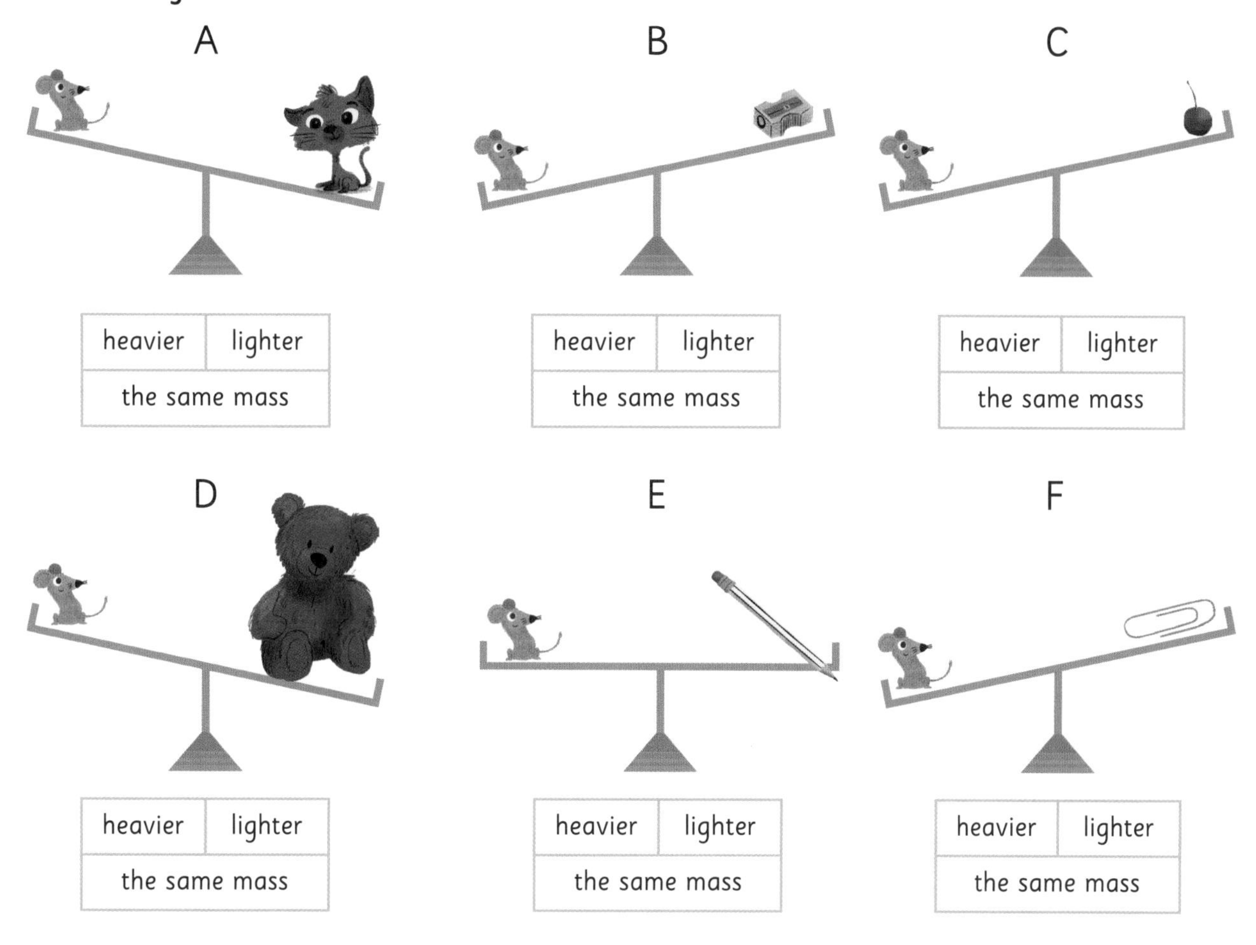

2 Order by mass from lightest to heaviest.

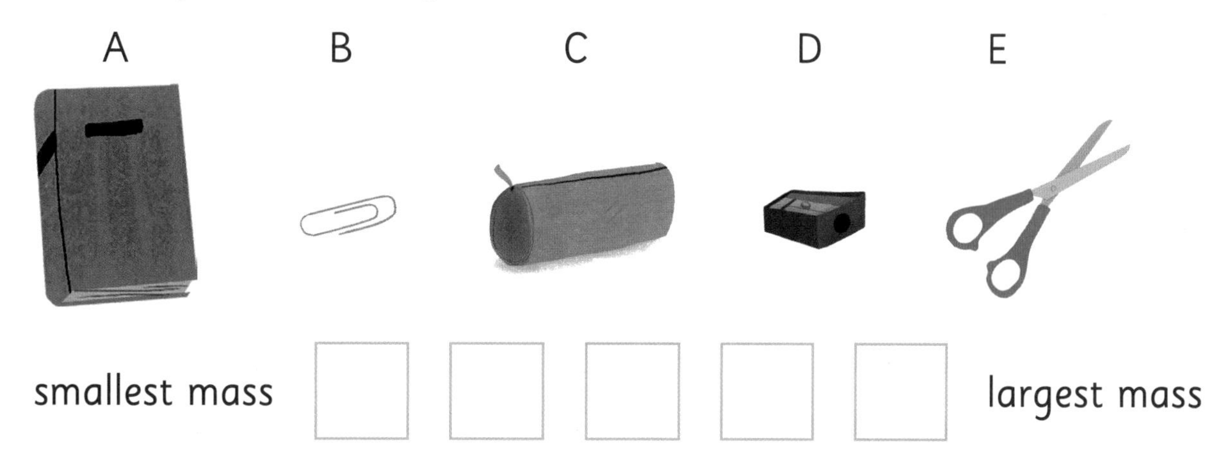

Challenge

1 Draw the objects at the correct end of the balance scale. Be ready to explain how you chose the heavier object.

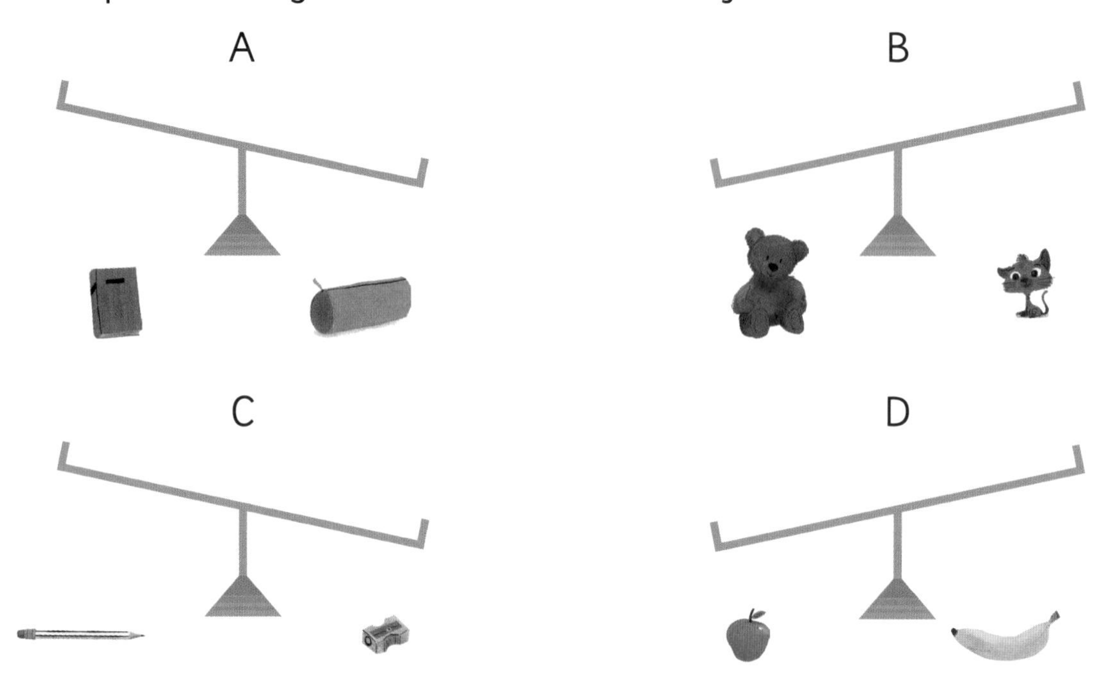

2 You will need a balance scale, a ruler and a pencil.

a Hold the ruler in one hand and the pencil in the other hand.

Which do you think is heavier? ____________________

b Put both items on the balance scale.

Which was heavier? ____________________

3 You will need a balance scale, a sharpener and an eraser.

a Hold the sharpener in one hand and the eraser in the other hand.

Which do you think is heavier? ____________________

b Put both items on the balance scale.

Which was heavier? ____________________

Mastery

You will need a balance scale and some beads or marbles in a box.

1 Put 10 beads or marbles on one side of the balance scale.

Find an object that balances them.

Draw the object.

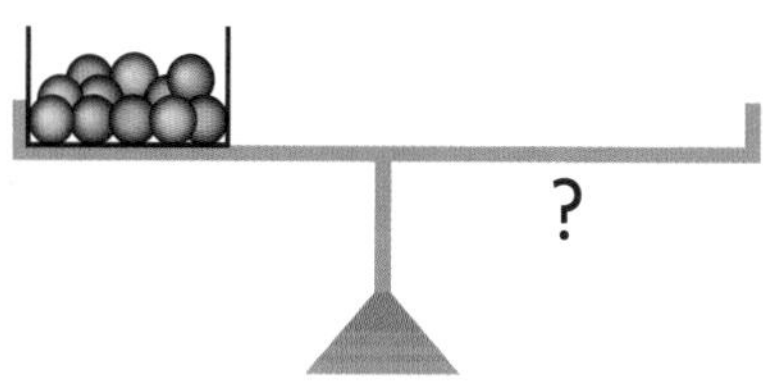

2 Put 20 beads or marbles on one side of the balance scale.

Find an object that balances them.

Draw the object.

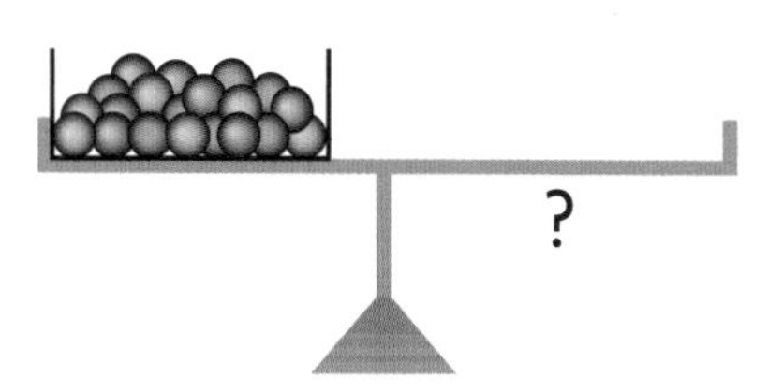

3 Place an object on one side of the balance scale.

Choose an appropriate unit to balance your object.

Draw the object on the scale and write how many of your unit you needed to balance it.

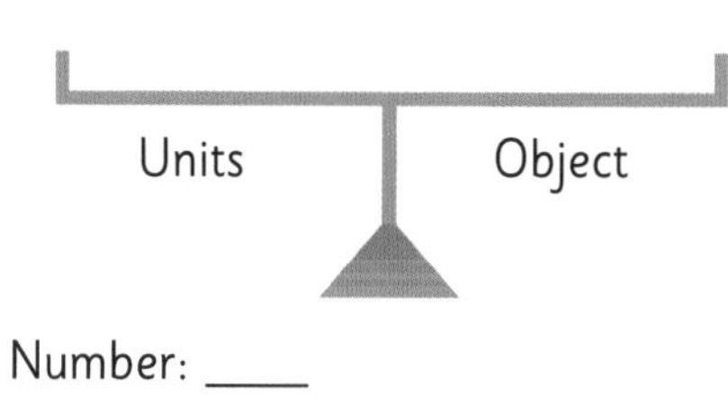

4 Repeat question 3 using a different object.

Use the same unit.

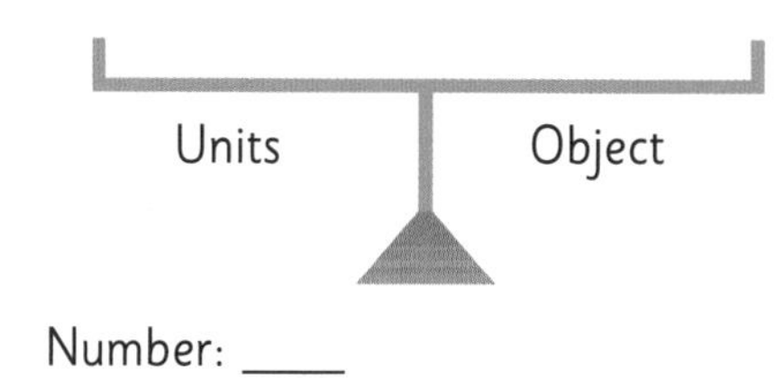

5

a Which of the objects in question 3 and 4 was heavier?

b By how much?

Time

Practice

1 Draw in the minute hands.

a half past 1

b quarter to 5

c quarter past 11

2 Draw in the hour hands.

a half past 7

b quarter to 4

c quarter past 2

3 Draw the times on the clocks.

a quarter past 7

b half past 3

c quarter to 8

4 Write in the times.

a

b

c

Challenge

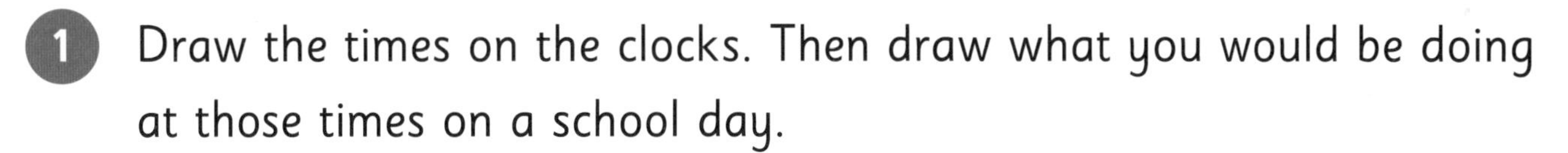

1 Draw the times on the clocks. Then draw what you would be doing at those times on a school day.

a half past 9 in the morning

b quarter past 5 in the morning

c quarter to 9 in the morning

d quarter past 1 in the afternoon

e half past 4 in the afternoon

f quarter to 10 in the evening

2

a Draw a time on the clock.

b Write in the time.

c Draw or write what you would be doing at that time during the weekend.

Time: ____________________

Mastery

This time can be written as quarter to 2 or 1:45.

1 Show each time in three ways.

a

quarter to 6

:

b

half past 8

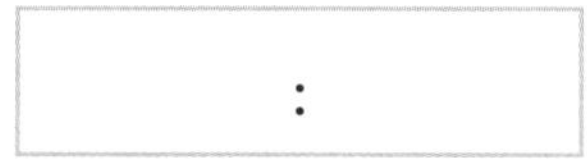

:

c

4:15

d

half past 6

:

e

3:45

f

quarter past 5

:

---

2 A movie starts at 1:45 and lasts for one and a half hours.

Show the start and end times in three different ways.

a Movie starts

:

b Movie ends

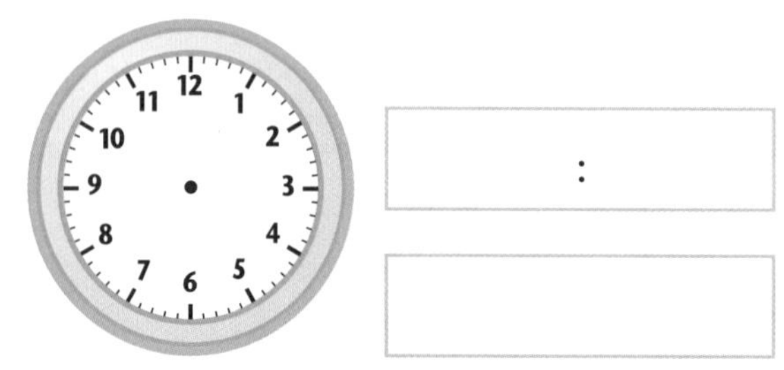

:

Calendars

Practice

1

JANUARY						
Sun	Mon	Tues	Wed	Thurs	Fri	Sat
	1	2	3	4	5	6
7	8	9	10	11	12	13
14	15	16	17	18	19	20
21	22	23	24	25	26	27
28	29	30	31			

FEBRUARY						
Sun	Mon	Tues	Wed	Thurs	Fri	Sat
				1	2	3
4	5	6	7	8	9	10
11	12	13	14	15	16	17
18	19	20	21	22	23	24
25	26	27	28			

MARCH						
Sun	Mon	Tues	Wed	Thurs	Fri	Sat
				1	2	3
4	5	6	7	8	9	10
11	12	13	14	15	16	17
18	19	20	21	22	23	24
25	26	27	28	29	30	31

APRIL						
Sun	Mon	Tues	Wed	Thurs	Fri	Sat
1	2	3	4	5	6	7
8	9	10	11	12	13	14
15	16	17	18	19	20	21
22	23	24	25	26	27	28
29	30					

MAY						
Sun	Mon	Tues	Wed	Thurs	Fri	Sat
		1	2	3	4	5
6	7	8	9	10	11	12
13	14	15	16	17	18	19
20	21	22	23	24	25	26
27	28	29	30	31		

JUNE						
Sun	Mon	Tues	Wed	Thurs	Fri	Sat
					1	2
3	4	5	6	7	8	9
10	11	12	13	14	15	16
17	18	19	20	21	22	23
24	25	26	27	28	29	30

JULY						
Sun	Mon	Tues	Wed	Thurs	Fri	Sat
1	2	3	4	5	6	7
8	9	10	11	12	13	14
15	16	17	18	19	20	21
22	23	24	25	26	27	28
29	30	31				

AUGUST						
Sun	Mon	Tues	Wed	Thurs	Fri	Sat
			1	2	3	4
5	6	7	8	9	10	11
12	13	14	15	16	17	18
19	20	21	22	23	24	25
26	27	28	29	30	31	

SEPTEMBER						
Sun	Mon	Tues	Wed	Thurs	Fri	Sat
						1
2	3	4	5	6	7	8
9	10	11	12	13	14	15
16	17	18	19	20	21	22
23	24	25	26	27	28	29
30						

OCTOBER						
Sun	Mon	Tues	Wed	Thurs	Fri	Sat
	1	2	3	4	5	6
7	8	9	10	11	12	13
14	15	16	17	18	19	20
21	22	23	24	25	26	27
28	29	30	31			

NOVEMBER						
Sun	Mon	Tues	Wed	Thurs	Fri	Sat
				1	2	3
4	5	6	7	8	9	10
11	12	13	14	15	16	17
18	19	20	21	22	23	24
25	26	27	28	29	30	

DECEMBER						
Sun	Mon	Tues	Wed	Thurs	Fri	Sat
						1
2	3	4	5	6	7	8
9	10	11	12	13	14	15
16	17	18	19	20	21	22
23	24	25	26	27	28	29
30	31					

Use this calendar to answer the questions.

a How many months have 31 days?

b How many months have 30 days?

c Which is the shortest month?

It has days.

d How many Sundays are there in July?

e If today is 15 August, how many days are left in the year?

Show how you worked out the answer.

Challenge

1 Use the calendar on page 59.

a Is 12 May a school day on the calendar? How do you know?

b Anzac Day is 25 April. What day is it on this calendar?

c What day and date is it one week after Anzac Day on the calendar?

d Which months are four weeks and two days long?

2 Choose a date that is special to you. Write the day and date and explain or draw why it is special.

3 This calendar shows the dates in one month of the year.

Sun	Mon	Tues	Wed	Thurs	Fri	Sat
		1	2	3	4	5
6	7	8	9	10	11	12
13	14	15	16	17	18	19
20	21	22	23	24	25	26
27	28	29	30	31		

a Could this be the month of June? How do you know?

b Which days are there five of in the month?

Mastery

1 This is part of a calendar for 2020.

a What is the difference between 1 February 2020 and 1 February on the calendar on page 59?

FEBRUARY 2020						
Sun	Mon	Tues	Wed	Thurs	Fri	Sat
						1
2	3	4	5	6	7	8
9	10	11	12	13	14	15
16	17	18	19	20	21	22
23	24	25	26	27	28	29

____

b What is the difference between the number of days in February 2020 and the number of days in February on the calendar on page 59?

____

2

a Write the month of your birthday at the top of the calendar.

b On what day is the 1st of your birthday month this year?

____

c Write 1 in the correct place on the calendar.

Month:						
Sun	Mon	Tues	Wed	Thurs	Fri	Sat

d Fill in all the other dates for your birthday month.

e Put a circle around the date of your birthday.

Practice

1

a Colour the shapes with 4 straight sides red.

b Colour the shapes with 3 corners blue.

c Colour the shapes with curved sides green.

2 Match the shapes to their names and descriptions.

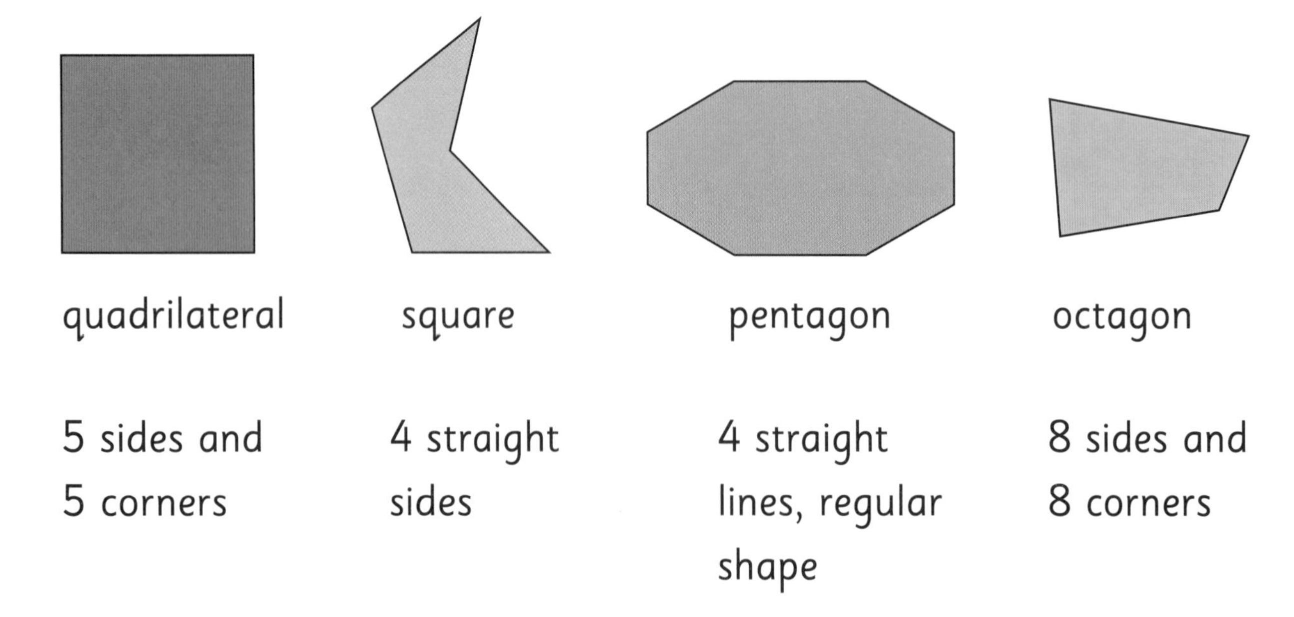

quadrilateral	square	pentagon	octagon
5 sides and 5 corners	4 straight sides	4 straight lines, regular shape	8 sides and 8 corners

Challenge

1 Draw the following shapes. Then circle whether the shape is regular or irregular.

a 6 sides and 6 corners

Regular Irregular

b 1 straight side and 1 curved side

Regular Irregular

c 4 straight sides all the same length

Regular Irregular

d 4 straight lines with two sets of parallel lines

Regular Irregular

---

2 Look at the shapes.

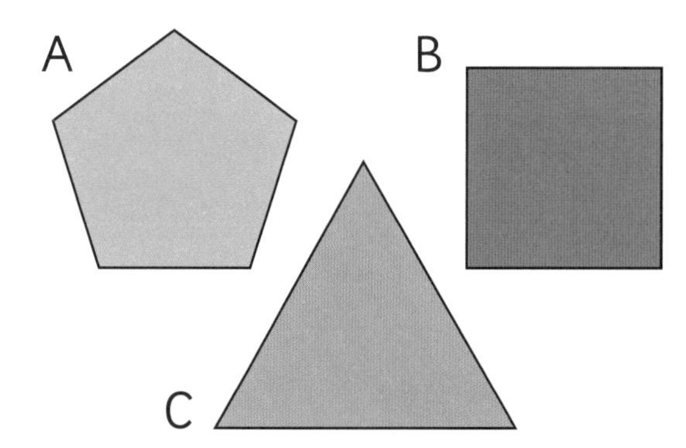

a What is the same about the three shapes?

____

____

b Apart from the colours, what is different about the shapes?

____

____

Mastery

1. Describe the shapes that have been used to draw this face.
   Use the shape names if possible.

2. Draw your own picture using 2D shapes. It does not have to be a face, but it needs to have some straight lines, some curved lines, some parallel lines, some opposite sides, some regular shapes and some irregular shapes. Colour it in.

   Describe what shapes you have used.

Practice

Use the word bank to help you.

triangular prism	sphere	square pyramid

1 This is a ______________.

It has:

a ☐ faces

b ☐ edges

c ☐ corners.

2 This is a ______________.

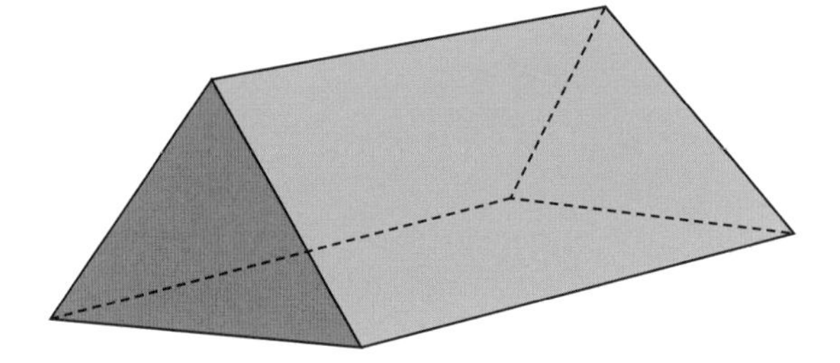

It has:

a ☐ faces

b ☐ edges

c ☐ corners.

3 This is a ______________.

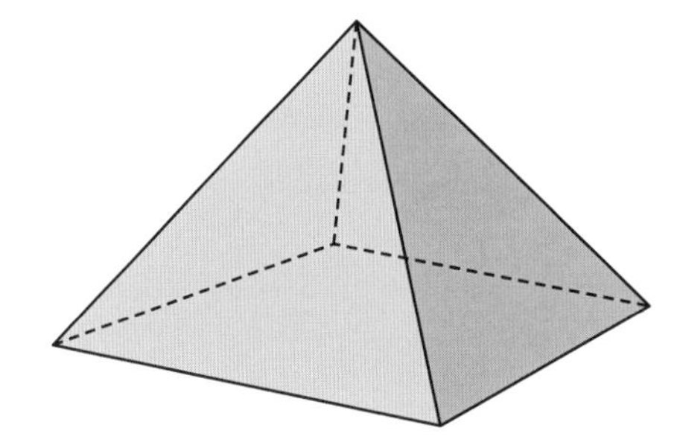

It has:

a ☐ faces

b ☐ edges

c ☐ corners.

4 Shade the object that has 1 flat face and 1 curved face.

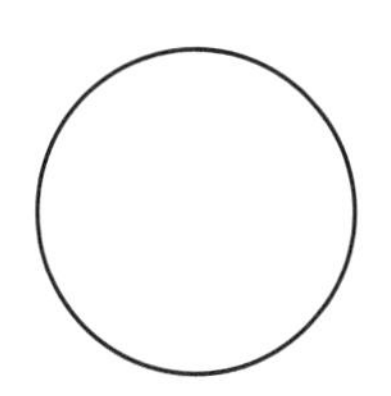

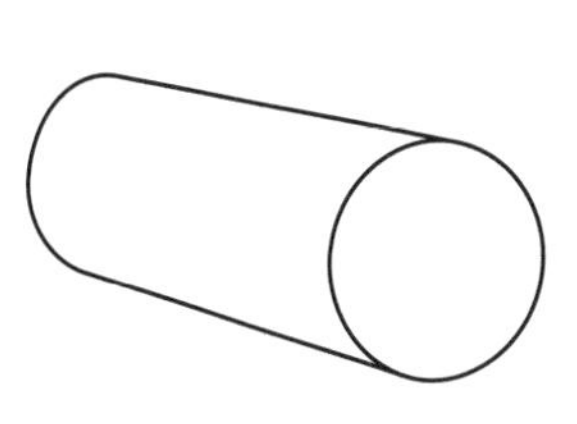

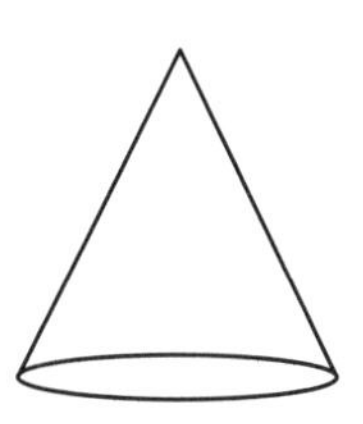

Challenge

1 Draw lines to match the objects with their names and descriptions.

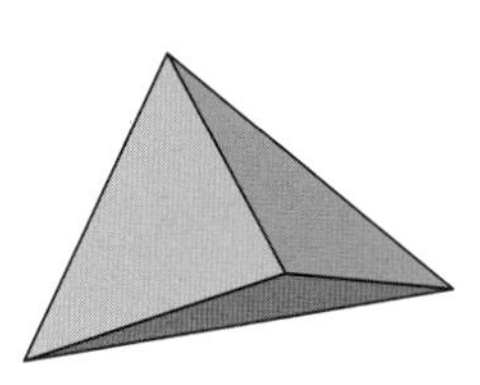 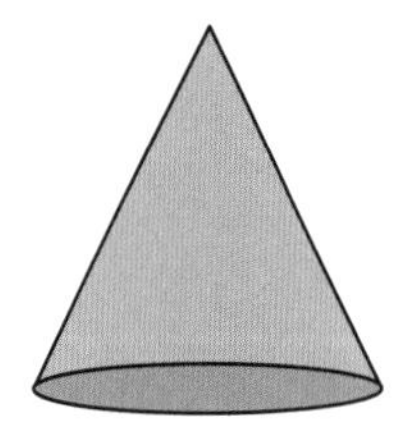 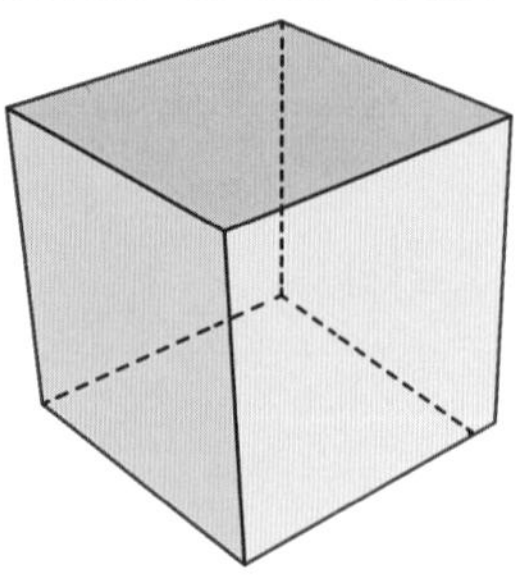 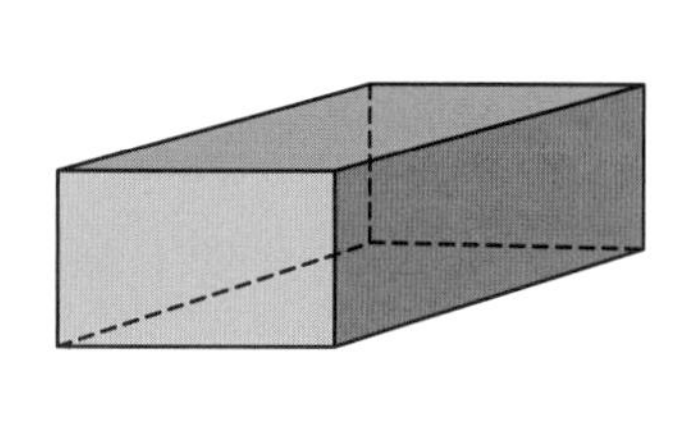

cone	rectangular prism	triangular pyramid	cube
1 flat face and 1 curved face	4 flat faces, 4 corners, 6 edges	12 edges, 8 corners, 6 rectangular faces	12 edges, 8 corners, 6 square faces

2 What am I?

a I have 4 faces.

My faces are triangles.

I am a ______.

b I have 3 faces.

2 of my faces are circles.

I have a curved face.

I am a .

c I have 5 faces.

4 of my faces are triangles.

One of my faces is a square.

I am a ______.

d I have 6 faces.

All of my faces are squares.

I am a .

Mastery

1 Join the dots to make 3D objects.

a

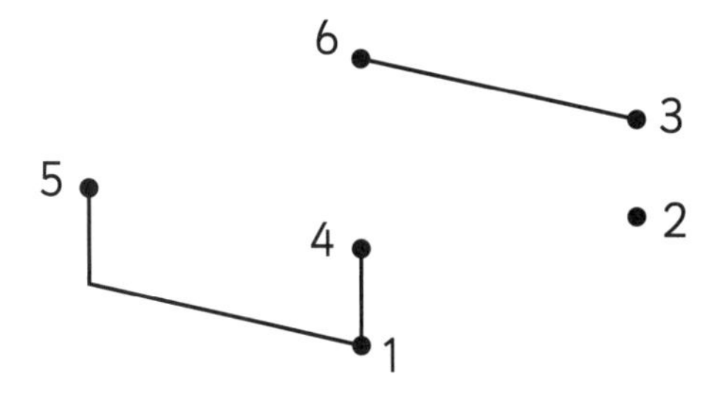

I have drawn a

b

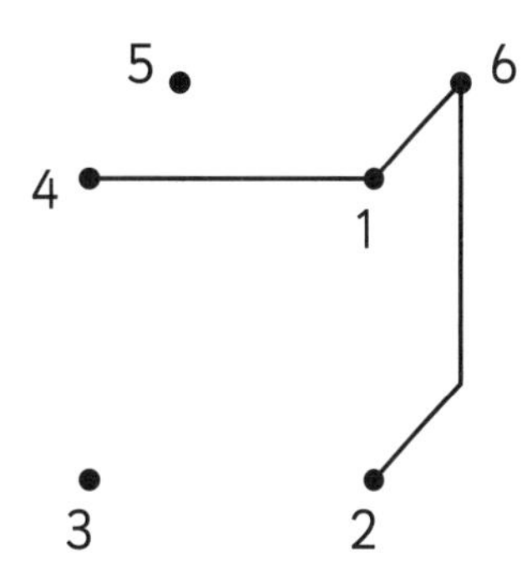

I have drawn a

---

2 Draw the whole object as follows:

a Join the dots from 1 to 5.

b Join the dots from a to c.

c Join 1 to a, 2 to b and 3 to c.

d What did you draw?

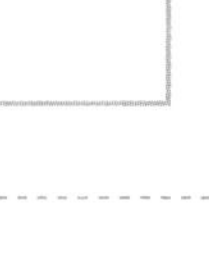

3

a Draw the object by joining the dots.

What did you draw?

b Draw the object on a separate piece of paper without using dots.

Interpreting maps and directions

Practice

1 Ava has a shelf for toys.

a The teddy is on the top shelf on the right. Where is the doll?

b What is below the doll?

c Where is the car?

d Where is the dinosaur?

2 A plan of something is like looking at it from above. This is a plan of Ava's room.

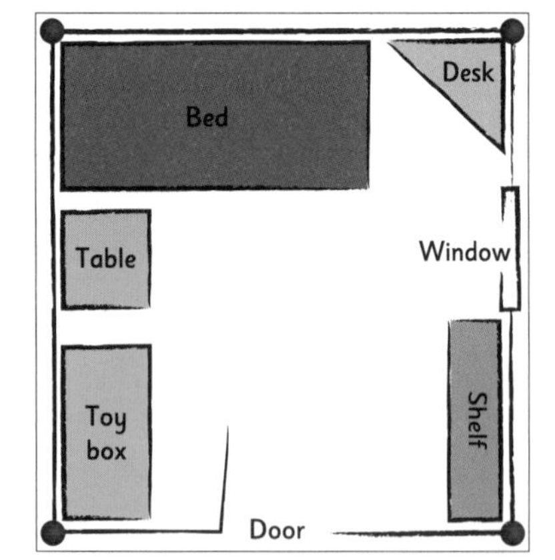

a What is between the toy box and the bed?

b What is opposite the table?

c Describe the position of the desk.

d Draw a green rug on the floor.
Describe its position.

Challenge

1

a Draw one of your favourite things in each box.

b What did you draw in the top right-hand box?

c What did you draw in the middle box at the bottom?

d Describe the positions of two other objects.

e Write a question about the position of one of the other objects.

2

a Draw a dog under the table between the table legs.

b Draw a friend for big teddy next to him.

c Draw a cat under the right-hand end of the table.

d Draw a butterfly above small teddy's head.

Mastery

1 Sometimes you need to use the words *row* and *column* to describe position.
A column goes up and down and a row goes across.
The star is in the 1st column on the 1st row.

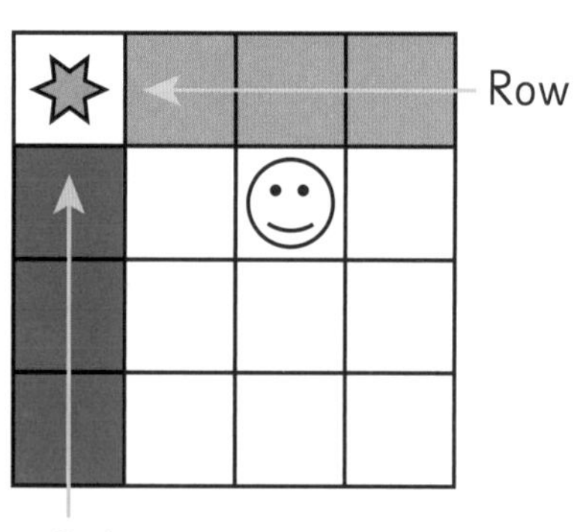

a Where is the smiley face?

b Draw another smiley face in the 4th column on the 3rd row.

c Draw another star in the 2nd column on the 4th row.

---

2 Find the hidden message by putting the letters in the boxes.

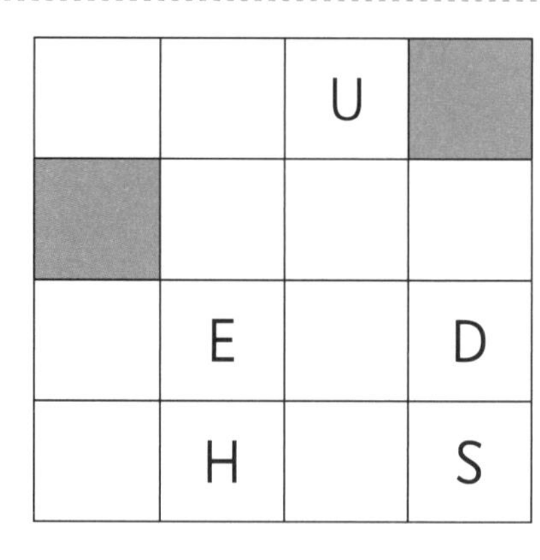

a Y in the 1st column on the 1st row

b R in the 1st column on the 3rd row

c N in the 4th column on the 2nd row

d C in the 2nd column of the 2nd row

e T in the 1st column on the bottom row

f A in the 3rd column on the 2nd row and 3rd column on the 3rd row

g O on the top row in the 2nd column

h Which white square has no letter?

i Write a letter in the white square to finish the sentence.

3

a Write the message "I AM OK" on these three rows.

b On a separate piece of paper, write instructions to tell someone where to write the letters.

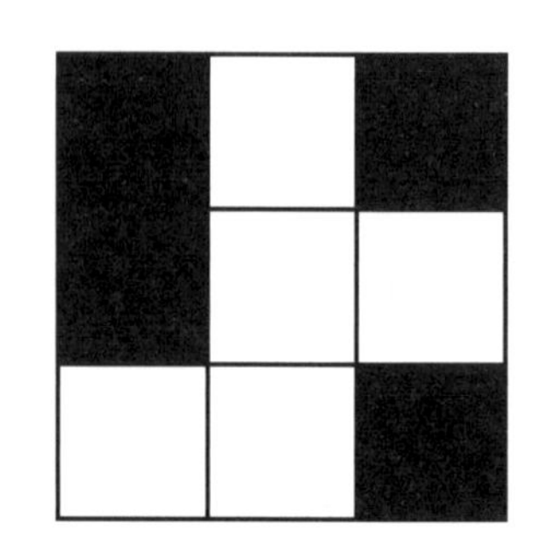

UNIT 7: TOPIC 2
Measures of turn

Practice

1 Decide whether the pattern is showing half turns or quarter turns, then continue the pattern.

a

Half turn	Quarter turn

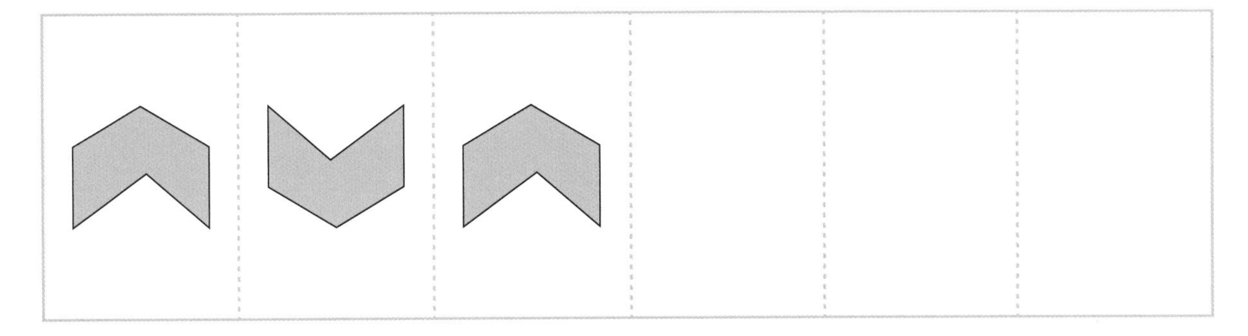

b

Half turn	Quarter turn

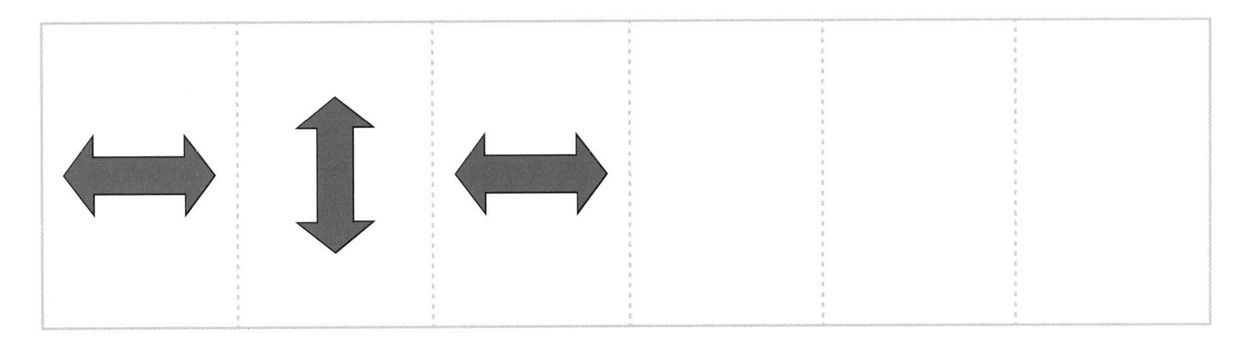

2 Draw the shapes after these turns.

a

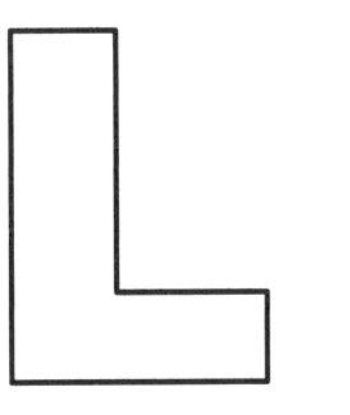

quarter turn

full turn

b

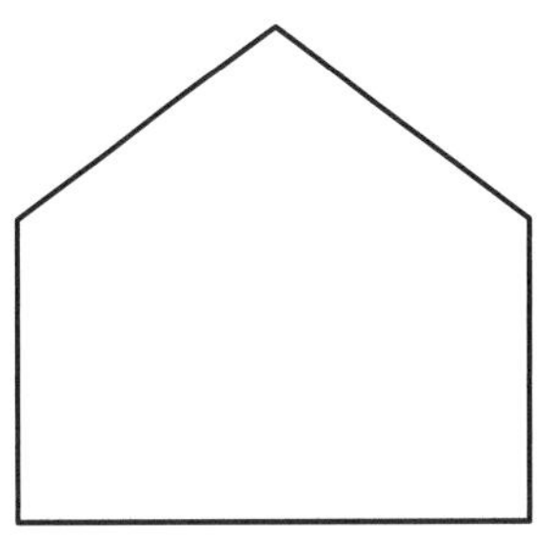

three-quarter turn half turn

Challenge

1 a Describe the turn used to make the pattern.

b Continue the pattern. Use colours if you wish.

2 a Describe the turn used to make the pattern.

b Continue the pattern. Use colours if you wish.

3 Look at the pattern to decide which turns have been used.

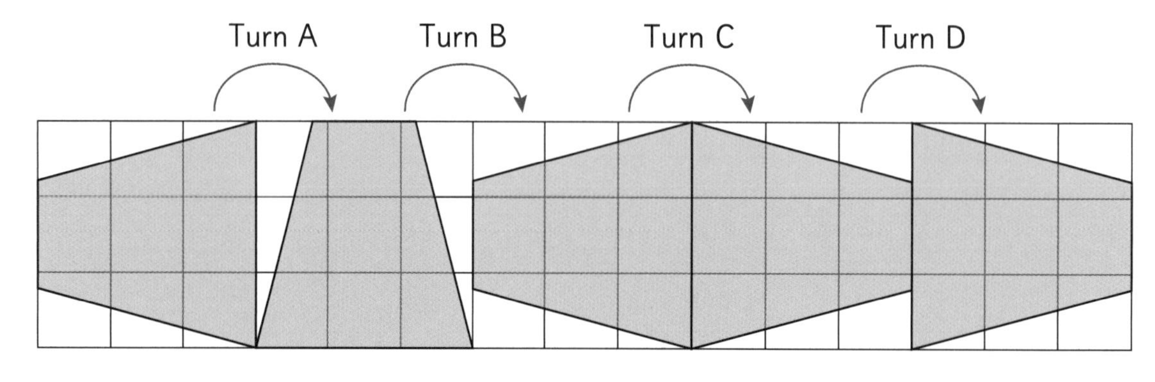

Turn A:

Turn B:

Turn C:

Turn D:

Mastery

1. There are many examples of turns in real life.

We turn on taps.

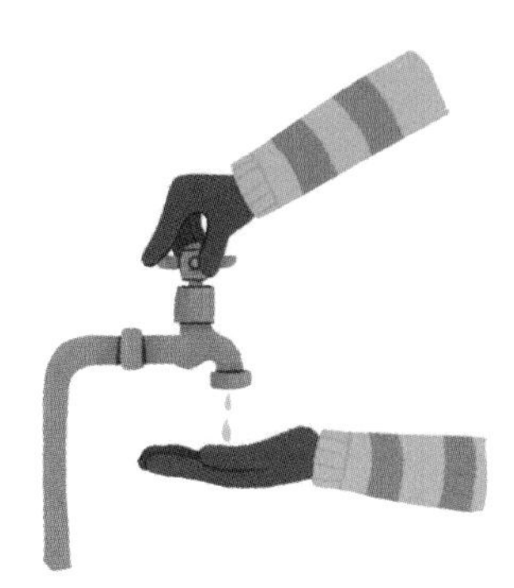

We turn corners.

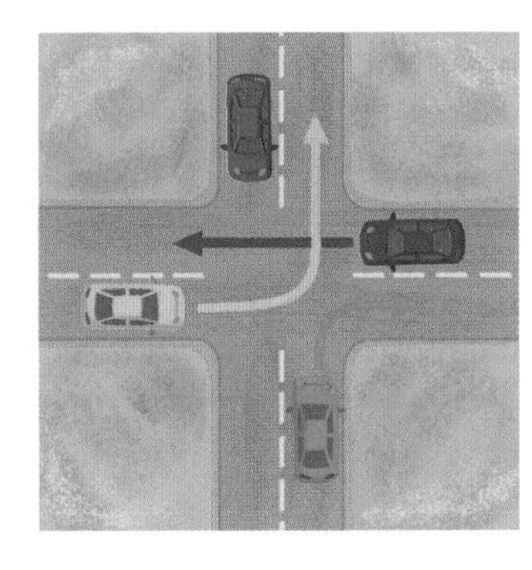

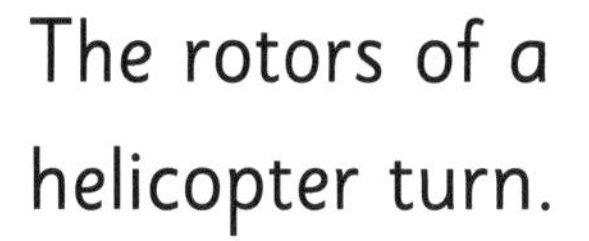

The rotors of a helicopter turn.

Draw or list where you have seen each of the turn types below in real life.

a quarter turns

b half turns

c three-quarter turns

d full turns

UNIT 8: TOPIC 1
Collecting data

Practice

1 Year 2 students wrote their names on the whiteboard.

Fruit	Students
Apple	Ava Jack Ben Sam Eve Tran Henry Finn Jo Emma
Orange	Nguyen Brad Joe Rafferty Owen Joel Audrey Susie
Strawberry	Hannah Kate Katie Zac Dee Ahmed
Banana	Sophia Jackson Noah Liam Olivia Poppy Lily George Mohammed Leo Freddie Max
Other fruit	Edward Riley Lottie Mollie Louis Luca Elijah Alfie Oliver Amelia

a Count the names and record them.

Fruit	Tick for each student	Total
Apple		
Orange		
Strawberry		
Banana		
Other fruit		

b What question do you think the teacher asked them?

c How many students are in Year 2?

Challenge

1

a Write a yes/no question about pets.

b Ask 12 students and record their answers.

Yes	No

2 Collect data from other students about food.

a What question will you ask?

b Record the data in this table.

Food	Responses

c Write a sentence about what the data shows.

Mastery

We can ask two types of questions to collect data: *closed questions* and *open questions.*

A yes/no question is a closed question. This is because there are only two ways to answer: yes or no. For example, we could ask, “Do you like sport?”

In an open question there are more than two ways to answer. We could ask, “What sport do you like to play?”

1. Write a different closed question to ask other students about sport.

2. Ask 12 or more students your closed question and record their answers.

Yes	No

3. Write an open question to ask other students about music.

4. Ask your open question to 12 or more students. Record their answers in a table like this on another sheet of paper.

Write your question here in the space at the top	
Names in this column	**Replies in this column**

Collecting and classifying data

Practice

1 12 students were asked the question "What is your favourite meal?"

Their answers were: lunch, breakfast, lunch, dinner, breakfast, lunch, brunch, lunch, dinner, dinner, dinner, dinner

Use this data to fill in the table below.

Meal	Breakfast	Brunch	Lunch	Dinner
Tally				
Total				

2

a Record two more subjects in the table.

Favourite subjects in our class				
Subject	Maths	Reading		
Tally				
Total				

b Survey 12 or more people in your class about their favourite subjects. Put tally marks in the table.

c Total the tallies.

d Which subject was the most popular?

e Which subject was the least popular?

Challenge

1

a Choose a way to sort the faces and write the categories at the top of the table.

	Categories			
Total				

b Use tally marks to record the number for each category in the table.

c Write a piece of information that your data shows.

d Write a question for someone else to answer about the data.

Mastery

1. Choose a topic that can be used to collect data, such as the type of TV programs that people watch or what people read. Then carry out a survey to collect the data. Use the space below to make a table and record the data. Write some information that your data shows.

OR

2. Use the monster picture to collect data.

a Choose a way to sort the monsters so that you can collect data about them.

b Create a table. Record a tally and find the total for each category.

c On a separate piece of paper, write some information that your data shows.

Representing and interpreting data

Practice

1 A Year 2 class collected data about pets.

a How many people have a dog? ☐

b Which pet is there least of? ☐

c How many more people have dogs than birds? ☐

d What type of pet might be "Other"? ☐

The pets our class has

10				
9				
8				
7				
6				
5				
4				
3				
2				
1				
	Dog	**Cat**	**Bird**	**Other**

2 **Length of hair in Year 2 class**

Hair length	Tally	Total
Long	卌 ||||	
Short	卌 ||	
Medium	卌 |||	

a Write the totals in the table.

b Finish the pictograph.

c Which is the most common hair length? ☐

9			
8			
7			
6			
5			
4			
3			
2			
1			
	Long		

Challenge

1 Hair length in our class

a Collect data about hair length in your class.

Hair length	Tally	Total
Long		
Short		
Medium		

b Make a pictograph of the results.

10			
9			
8			
7			
6			
5			
4			
3			
2			
1			

c What is the most common hair length in your class?

[ ]

2 Year 2 did a survey about what people like to do at playtime.

Key: ☺ = 1 person

Handball	Running around	Football	Talking with friends
☺☺☺☺☺☺☺☺	☺☺☺☺☺	☺☺☺☺☺	☺☺☺☺☺☺☺

a What do most people in this Year 2 like to do at playtime?

[ ]

b How many people are in this Year 2? [ ]

c Which two activities are as popular as each other?

[ ] [ ]

Mastery

1 What do people like to do at the weekend?

To find out, you need to think of how many categories there might be.

If you ask, "What do you like to do at the weekend?" you might get more than 20 different answers. That would be too many.

a Begin by writing no more than six categories of activity in the table. You could start with *Play sport*, *Read*, and *Play with friends*.

Activity	Tally	Total

b Write your survey question based on your categories.

c Carry out the survey using tally marks or ticks. You may wish to discuss with your teacher who you should ask.

d Write the totals in the table.

e Use the data to make a pictograph similar to the ones in this topic. Do this on a separate piece of paper.

f Write a sentence or two about the information shown in your survey.

Chance

Practice

1 Choose the word that matches each situation.

Impossible	Unlikely	Likely	Certain

a It is somebody's birthday today.

b You will travel in a plane today.

c A cow will jump over the moon.

d You will drink water today.

2 Draw or write something that is:

a unlikely to happen.

b likely to happen.

Challenge

1. Imagine choosing a balloon while you are blindfolded. Match each description to a picture by drawing a line.

a impossible to pick a yellow balloon

b likely to pick a yellow balloon

c unlikely to pick a yellow balloon

d certain to pick a yellow balloon

2. Describe the chance of picking a purple flower without looking.

a

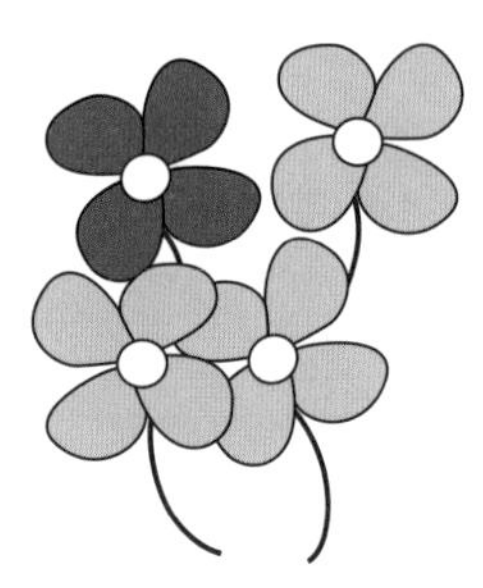

b

c

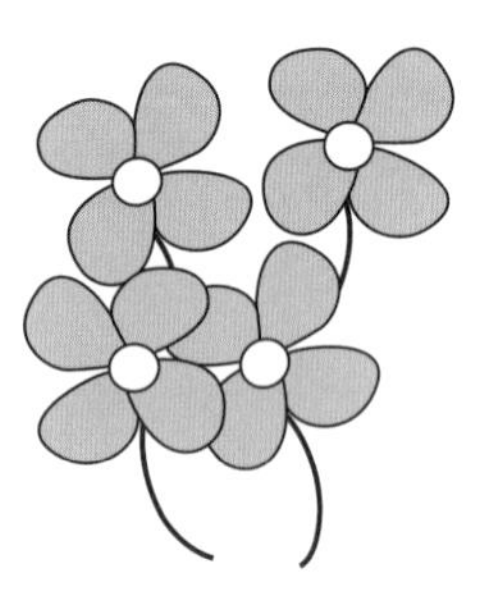

d

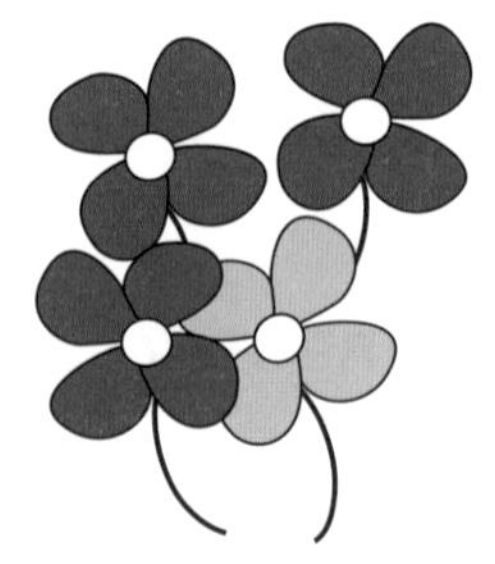

3. Draw or write something that is likely to happen tomorrow.

Mastery

1 A blindfold chance game

Imagine a friend choosing a red bead without looking.

Using only red and yellow, colour the beads so that the chance of choosing a red bead is:

a likely. b certain. c impossible. d unlikely.

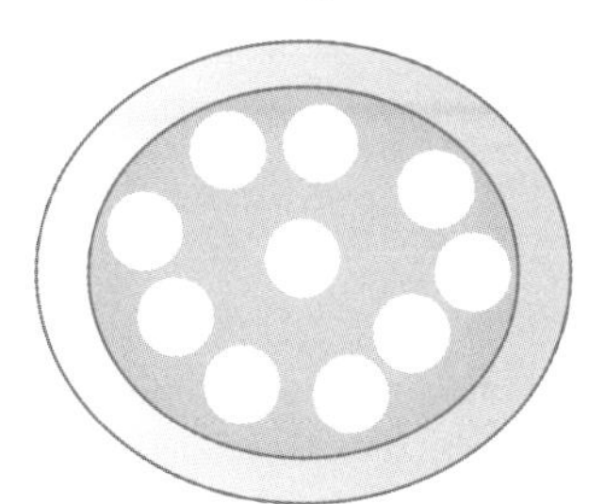

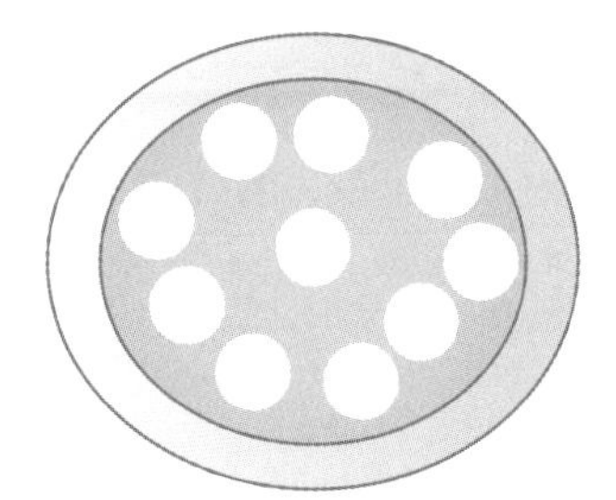

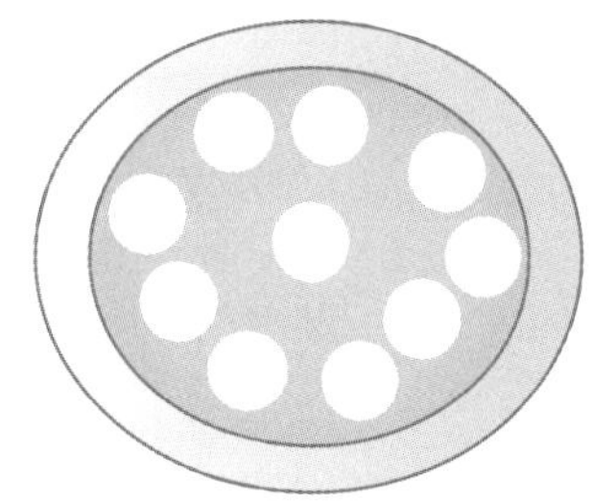

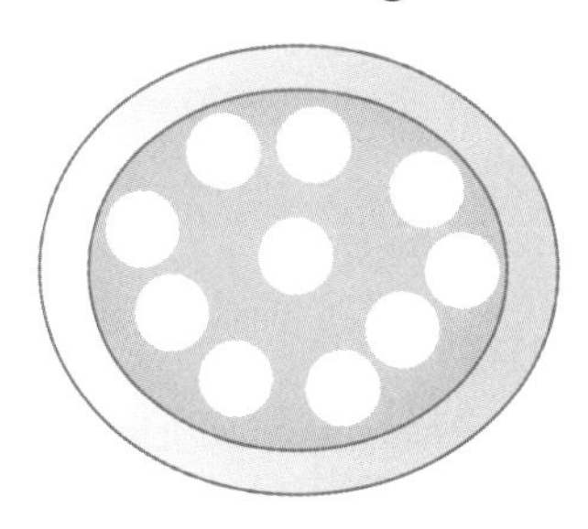

2

Read the instructions carefully.

There are 20 beads on this plate.

Colour the beads red, yellow and blue so that:

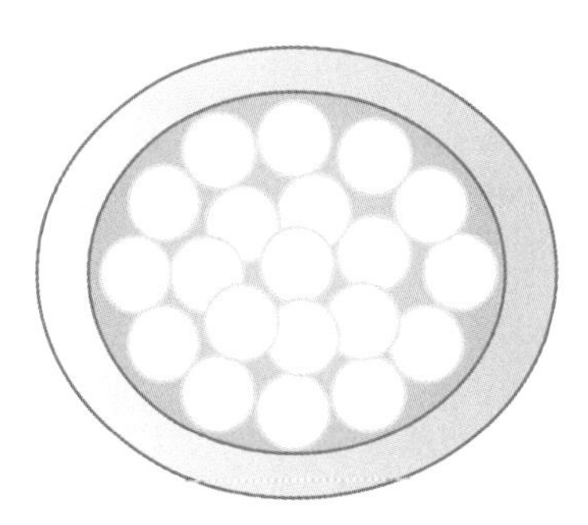

a the chance of choosing a blue bead or a yellow bead is unlikely.

b i the chance of choosing a blue bead is likely.

ii the chance of a red bead is unlikely.

iii the chance of a yellow bead is less likely than choosing a red bead.

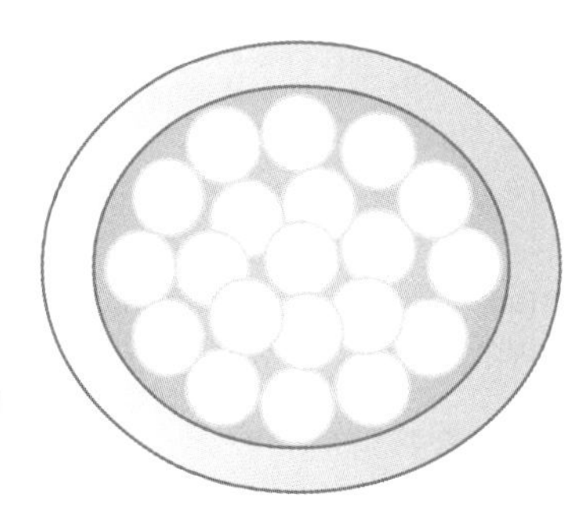

c i it is impossible to choose a red bead.

ii the chance for a blue bead is the same as the chance for a yellow bead.

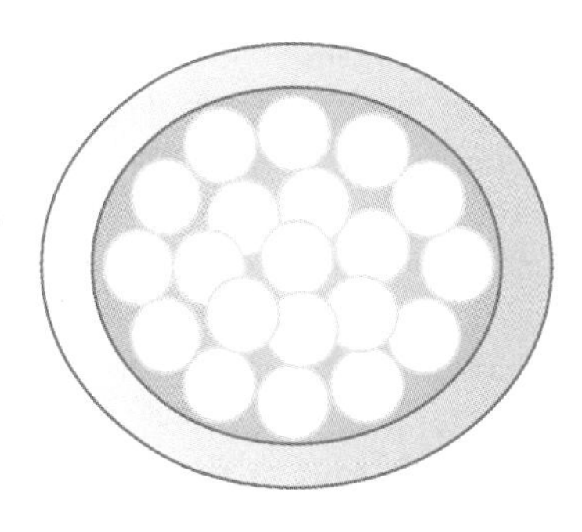

ANSWERS

UNIT 1: Topic 1

Practice

1 a 23 b 68 c 124 d 146

2 a 154 b 397 c 673

3 2 hundreds, 1 ten, 9 ones

Challenge

1 a 3 hundreds, 7 tens, 5 ones
  b 6 hundreds, 0 tens, 7 ones

2 a 70 b 800

3 a 359 b 935

4 Teacher to check.

Mastery

1 a–b Answers will vary but should follow the pattern of the hundred chart.

2 123, 132, 213, 231

3 There are six combinations:
  678, 687, 768, 786, 867, 876.

4 a 350
  b Teacher to check. Look for students who show an understanding of numeration, counting and place value in their explanations, regardless of the estimation.

UNIT 1: Topic 2

Practice

1 a 6 hundreds 7 tens 1 one
  b 5 hundreds 17 tens 1 one
  c 6 hundreds 5 tens 21 ones

2 a and b Teacher to check. Look for students who can accurately draw and write regroupings of the numbers.

Challenge

1 Draw lines to match the numbers that are the same.

5 hundreds, 14 tens and 7 ones	6 hundreds, 10 tens and 9 ones
3 hundreds and 48 ones	3 hundreds, 12 tens and 54 ones
7 hundreds and 9 ones	6 hundreds, 3 tens and 17 ones
3 hundreds, 15 tens and 8 ones	3 hundreds, 4 tens and 8 ones
4 hundreds, 2 tens and 54 ones	4 hundreds and 58 ones

2 a 298 29 tens and 8 ones or 298 ones
  b 501 50 tens and 1 one or 501 ones
  c 888 88 tens and 8 ones or 888 ones

3 Rename each number without using tens.
  a 298 2 hundreds and 98 ones or 298 ones
  b 501 5 hundreds and 1 one or 501 ones
  c 888 8 hundreds and 8 ones or 888 ones

Mastery

1 a CCCI b MCC c DCCLVIII
  d DXXV e DCCCXIV f CLXVII

2 Teacher to check. Look for students who demonstrate an understanding of how the base-10 system makes it easier to count and represent numbers. Students' responses may be used to inform further teaching of place value.

3 Answers may vary. For example:
  a 734 6 hundreds, 13 tens and 4 ones or 6 hundreds, 3 tens and 104 ones
  b 576 5 hundreds, 6 tens and 16 ones or 5 hundreds, 1 ten and 66 ones
  c 772 6 hundreds, 16 tens and 12 ones or 6 hundreds, 15 tens and 22 ones

UNIT 1: Topic 3

Practice

1 a 18 b 22 c 26
  d 42 e 60 f 88

2 a 10 + 11 = 10 + 10 + 1 = 20 + 1 = 21
  b 12 + 13 = 12 + 12 + 1 = 24 + 1 = 25
  c 15 + 16 = 15 + 15 + 1 = 30 + 1 = 31
  d 22 + 23 = 22 + 22 + 1 = 44 + 1 = 45
  e 41 + 44 = 41 + 41 + 3 = 82 + 3 = 85

3 46 + 8 = 46 + 4 + 4
  = 50 + 4
  = 54

Challenge

1 a 62 b 87 c 44
  d 69 e 100 f 110

2 a 44 b 61 c 52
  d 74 e 101 f 103

3 Answers may vary. For example:
  a 14 + 17 = 31 b 13 + 18 = 31
  c 12 + 19 = 31 d 11 + 20 = 31

4 $15 + $18 = $33

Mastery

1 The 10 ways are (in any order):
  a 40 + 12 = 52 b 41 + 11 = 52
  c 42 + 10 = 52 d 43 + 19 = 62
  e 44 + 18 = 62 f 45 + 17 = 62
  g 46 + 16 = 62 h 47 + 15 = 62
  i 48 + 14 = 62 j 49 + 13 = 62

2 There are still 10 different ways (in any order):
  a 40 + 40 = 80 b 41 + 49 = 90
  c 42 + 48 = 90 d 43 + 47 = 90
  e 44 + 46 = 90 f 45 + 45 = 90
  g 46 + 44 = 90 h 47 + 43 = 90
  i 48 + 42 = 90 j 49 + 41 = 90

3 There are always 10 ways to fill the boxes. For example, 3_ + 1_ = _9:
  a 30 + 19 = 49 b 31 + 18 = 49
  c 32 + 17 = 49 d 33 + 16 = 49
  e 34 + 15 = 49 f 35 + 14 = 49
  g 36 + 13 = 49 h 37 + 12 = 49
  i 38 + 11 = 49 j 39 + 10 = 49

UNIT 1: Topic 4

Practice

1 Look for students who start with the larger number.
  a 17 + 9 = 26

0 1 2 3 4 5 6 7 8 9 10 11 12 13 14 15 16 17 18 19 20 21 22 23 24 25 26 27 28 29 30

  b 34 + 11 = 45

OR

30 31 32 33 34 35 36 37 38 39 40 41 42 43 44 45 46 47 48 49 50

2 Look for students who change the order of the addends to make a simpler equation. For example:
  a 9 + 1 + 5 = 15
  b 6 + 4 + 8 = 18

3 Possible solution is:
  7 + 3 + 6 = 16

Challenge

1 Teacher to check. Look for students who show the ability to space their numbers accurately and correctly represent the addition problem, and those who show their level of understanding by rewriting the problem.
  24 + 7 = 31

2 Teacher to check. Look for answers that show students' ability to understand that they can partition numbers into 10s to add more easily, and for students who

use skip counting in their jumps rather than making steps of 1.

a 8 + 31 = 39

b 25 + 33 = 58

3 Answers may vary. Look for students who rewrite the problems in the first instance to make them easier to solve. For example:

a 6 + 4 + 7 = 17

b 17 + 13 + 12 = 42

Mastery

1 5

2 a The magic number is 18.

9	2	7
4	6	8
5	10	3

b The magic number is 30.

8	18	4
6	10	14
16	2	12

c The magic number is 45.

18	3	24
21	15	9
6	27	12

3 Look for students who recognise that the squares are variations of the square in question 1, in different orientations.

4	9	2
3	5	7
8	1	6

4	3	8
9	5	1
2	7	6

6	1	8
7	5	3
2	9	4

4 Students may need to research the solution or teachers could provide a few of the numbers in the grid.

8	11	14	1
13	2	7	12
3	16	9	6
10	5	4	15

UNIT 1: Topic 5

Practice

1 23 – 9 = 23 – 3 – 6
= 20 – 6
= 14

2 24 – 7 = 24 – 4 – 3
= 20 – 3
= 17

3 18 + 2 = 20
20 + 5 = 25
So, 25 – 18 = 7

4 12

Challenge

1 a 8 b 8
c 5 d 9

2 a 12 b 13
c 29 d 25

3 Teachers may choose to ask students to explain their strategies.
a $11 b 17 books c 18 stickers

Mastery

1 Answers may vary. For example:
The skateboard costs $48. The boy has $25. He needs $23 more.

2 Answers may vary but the subtraction fact should match the information. For example: Tess had 40 lollies. She gave away 17. She now has 23.
40 – 17 = 23

3 The bike in Shop A is $96 – $11 = $85. The bike in Shop B is $94 – $8 = $86. So, Shop A has the cheaper bike.

4 Solutions will vary. This could be carried out as a group activity. Look for students who use the solution to the first part of each to find an easy solution for the second part. Possible answers are:

a 32 – 6 = 26, 33 – 6 = 27

b 60 – 40 = 20, 61 – 41 = 20

UNIT 1: Topic 6

Practice

1 a 23 – 9 = 14
b 46 – 18 = 28

2 a 18 + 9 = 27
27 – 9 = 18
9 + 18 = 27
27 – 18 = 9
b 36 + 8 = 44
44 – 8 = 36
8 + 36 = 44
44 – 36 = 8

3 25 – 8 = 17

Challenge

1 a 24 – 7 = 17
b 49 – 23 = 26
c 34 – 5 – 4 = 25

2 a 9 + 17 = 26
26 – 9 = 17
17 + 9 = 26
26 – 17 = 9
b 27 + 15 = 42
42 – 15 = 27
15 + 27 = 42
42 – 27 = 15

Mastery

1 20

2 Answers will vary. The difference should always be 15.

3 Answers will vary. Teacher to check and could ask students to make up more than one number sentence that would be correct.

4 The difference between their heights is 93 – 28 = 65 cm. Students could share their strategies for finding the answer.

UNIT 1: Topic 7

Practice

1 a $5 \times 3 = 15$ b $3 \times 10 = 30$

2 a $6 \times 4 = 24$

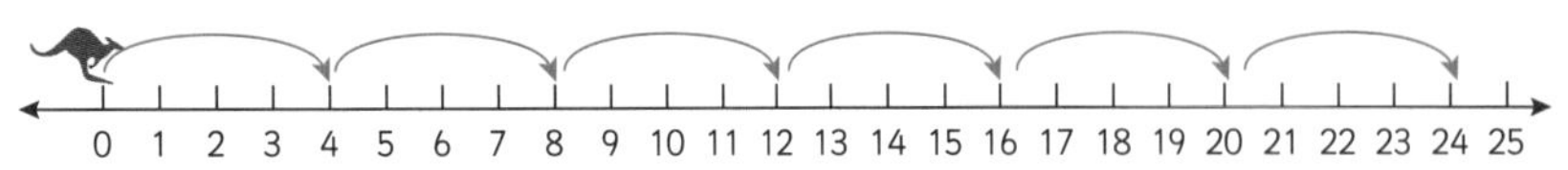

b $5 \times 5 = 25$

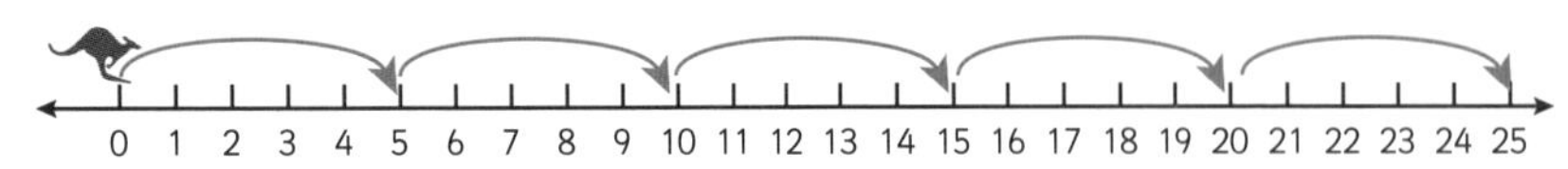

3 $3 \times 2 = 6$ $6 \times 2 = 12$

$9 \times 2 = 18$ $0 \times 2 = 0$

4 $2 \times 5 = 10$

Challenge

1 **a** Student draws 4 rows of 5 dots and writes $4 \times 5 = 20$.

b Student draws 5 rows of 4 dots and writes $5 \times 4 = 20$.

2 Student draws and writes $2 \times 4 = 8$, $4 \times 2 = 8$, $1 \times 8 = 8$.

3 **a** Yes.

b Teacher to check. Responses may vary, e.g. "... because she skips on 4 to 24 and then 4 to 28."

Mastery

1 Student draws and writes (in any order):

$6 \times 4 = 24$, $4 \times 6 = 24$, $2 \times 12 = 24$, $12 \times 2 = 24$, $3 \times 8 = 24$, $8 \times 3 = 24$

Students may also include $1 \times 24 = 24$ and $24 \times 1 = 24$

2

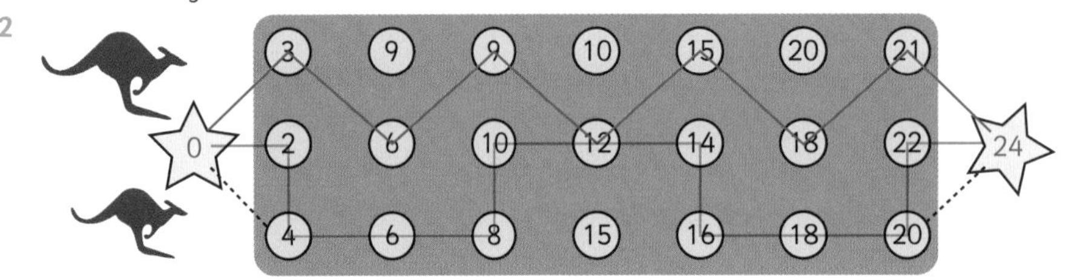

UNIT 1: Topic 8

Practice

1 a 10 b 2

c 10 divided by 2 = 5

$10 \div 2 = 5$

2 Student draws 2 flowers in each pot and writes:

10 divided by 5 = 2

$10 \div 5 = 2$

3 Student draws 4 beads on each plate and writes:

12 divided by 3 = 4

$12 \div 3 = 4$

4 Student shows 4 counters in each box and writes:

20 divided by 5 = 4

$20 \div 5 = 4$

Challenge

1 **a–b** Student makes groups of 3 and 5 and writes $15 \div 5 = 3$ and $15 \div 3 = 5$.

2 Students could make groups of 1, 2, 3, 4 or 12 stars. Likely answers are:

$12 \div 6 = 2$, $12 \div 3 = 4$, $12 \div 4 = 3$

3 12, 18, 24

Mastery

1 a Teacher to check. Answers may vary, e.g. "... because they can only share 15 between them equally."

b Answers may vary. This could be used as a group/class discussion point.

2 a–b 8 lollies should be drawn in each jar.

c $40 \div 5 = 8$

3 Practical activity. Students could be asked to write the arrays that they find.

UNIT 2: Topic 1

Practice

1 a $\frac{1}{4}$ b $\frac{1}{8}$

c $\frac{1}{2}$ d $\frac{1}{8}$

2 a

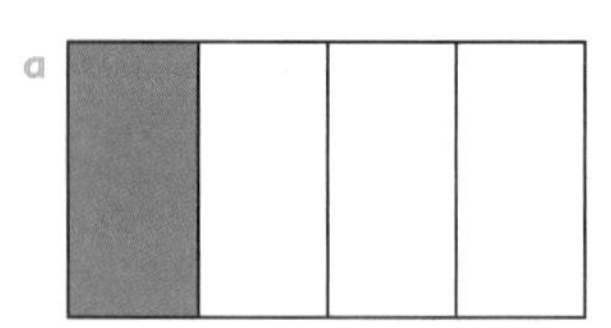

b

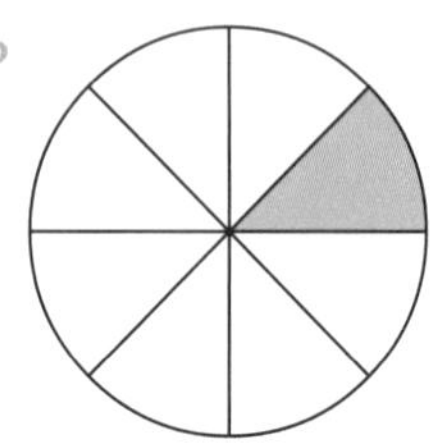

3

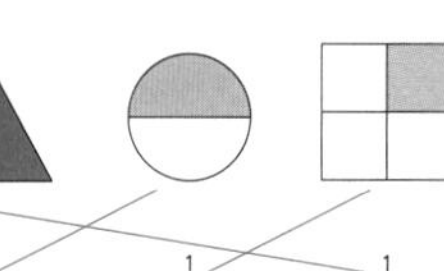

Challenge

1 Positions of sectors shaded may vary.

a–c

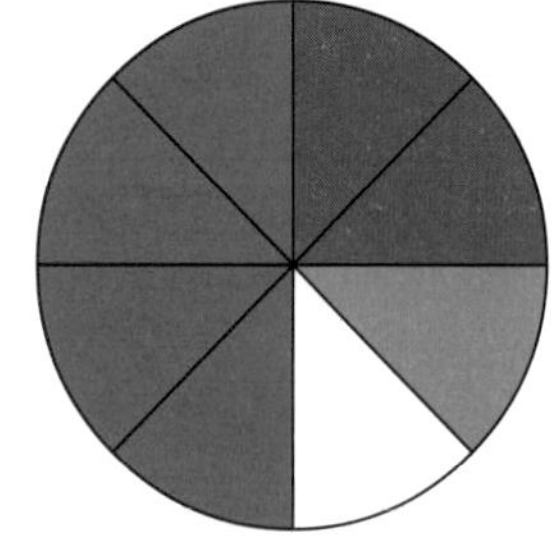

d $\frac{1}{8}$

2 **a–c** Teacher to check. Students may draw lines on the shapes to help them divide and show $\frac{1}{4}$s, $\frac{1}{8}$s, $\frac{1}{2}$s.

3 **a**

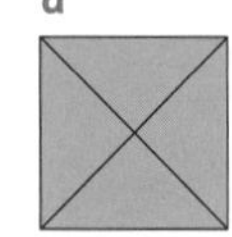
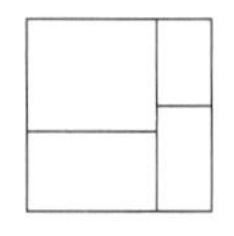
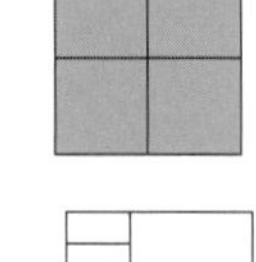
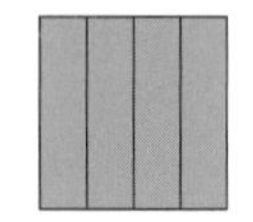
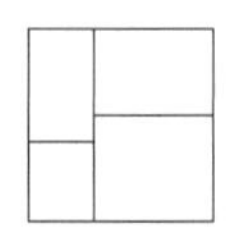

b Teacher to check. This could be part of a group discussion, e.g. "... because the pieces are not the same size."

Mastery

1 a–c Teacher to decide on expected level of accuracy. Look for answers that show a reasonable attempt to divide the bees. Possible answer is:

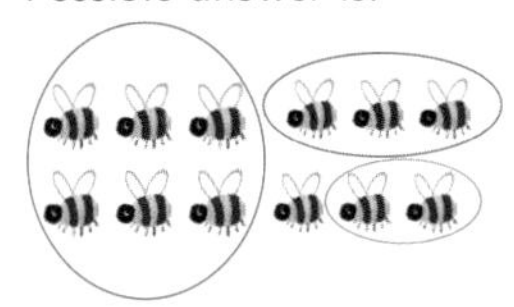

Group c is the smallest.

2 3 sections should be shaded.

3 3 sections should be shaded.

4 Practical activity. Teachers may wish to discuss options for shading half of each shape in an interesting way.

UNIT 2: Topic 2

Practice

1 a 2 items should be coloured in.

b 2 items should be coloured in.

2 a $\frac{1}{2}$ b $\frac{1}{4}$

3 a 2 items should be circled. $\frac{1}{4}$ of 8 is 2.

b 3 items should be circled. $\frac{1}{4}$ of 12 is 3.

4 a–c Some combination of the following:

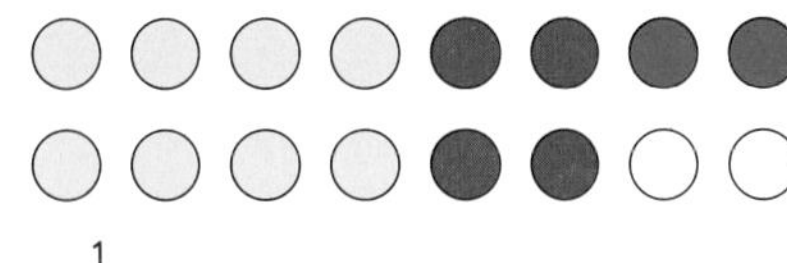

d $\frac{1}{8}$

Challenge

1 a Student draws hats on 5 faces.

b Student draws 4 smiley and 2 sad faces.

2 Student circles:

a $\frac{1}{2}$ of 10 b $\frac{1}{4}$ of 12

3 a $\frac{1}{4}$ b $\frac{1}{2}$ c $\frac{1}{8}$ d $\frac{1}{4}$

Mastery

1 Teacher to check. Possible answer is, "She would choose $\frac{1}{4}$ of 40 because that is 10, and $\frac{1}{2}$ of 16 is only 8."

2 Teacher to check. Students could share their ideas with each other.

3 a–b Teacher to check. Students could ask each other questions based on the tasks in this topic.

UNIT 3: Topic 1

Practice

1

How many of these ...	... do you need to make this?	Draw the answer	Write the answer
5	20	5 5 5 5	5c + 5c + 5c + 5c = 20c
10	50	10 10 10 10 10	10c + 10c + 10c + 10c + 10c = 50c
50	1 DOLLAR	50 50	50c + 50c = $1

2 a Student circles one $20, one $10 and one $5.

b Student circles three $20 and one $5.

3 a $49.50

b $36.20

Challenge

1 Student draws:

a three 5c coins

b two 20c and two 5c

one 20c and three 10c

c three $5 and one $10

d one $5, one $1, one 20c and one 5c

2 Student draws:

a one 5c, one 10c, one 20c, one 50c, one $1, one $2 and writes the total as $3.85

b one $5, one $10, one $20, one $50, one $100 and writes the total as $185

3 Possibilities are:

three 10c

two 10c and two 5c

one 10c and four 5c

six 5c

one 20c and one 10c

one 20c and two 5c

Mastery

1 Students could pool their ideas. Any combination of notes totalling $100 is acceptable.

2 Students could pool their ideas. Any combination of notes and coins totalling $20 is acceptable.

3 Student draws two $2 to pay and three 50c in change or one $1 and one $2 to pay and two 20c and one 10c in change.

UNIT 3: Topic 2

Practice

1 a $2.05

b $100

2 Students may draw or write their answers.

a three 20c coins

b twelve 5c coins

3 Students may draw or write their answers.

a four $50 notes

b ten $20 notes

Challenge

1 a 4 coins (one $1, one 50c, one 20c, one 10c)

b 7 coins (four $2, one 20c, one 10c, one 5c)

2 a $58.70

b i $38.70

ii $23.70

iii $5.70

3 a $77.25

b i $57.25 ii $8.25

iii $12.25

Mastery

1 a $3 b $6 c $10
 d $5 e $20 f $9

2 stapler: $3, eraser: 90c, pencil: 60c, sharpener: 70c

3 a 10c b $1.30 c $4.40
 d $17 e $6.40 f 70c

UNIT 4: Topic 1

Practice

1 a & c

10	20	30	40	50	60	70	80	90	100

b 10s

2 a & c

40	36	32	28	24	20	16	12	8	4

b 4s

3 a

3	**6**	9	12	**15**	**18**	21	**24**	**27**	**30**

b

1	**3**	5	7	**9**	11	**13**	**15**	**17**	**19**

c

51	46	41	**36**	**31**	**26**	21	**16**	**11**	**6**

d

101	**91**	81	**71**	61	**51**	**41**	**31**	**21**	**11**

Challenge

1 a 4s

b–c

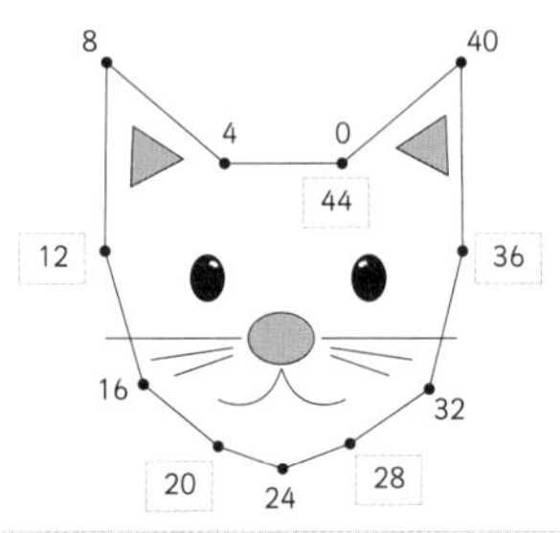

2 a–c

1	2	3	4	5	6	7	8	9	10
11	12	13	14	15	16	17	18	19	20
21	22	23	24	25	26	27	28	29	30
31	32	33	34	35	36	37	38	39	40

3 a 8 b 4
 c 16 d 32

Mastery

1 a–i Teacher to check. Students could share their findings with each other. Teachers may wish to guide students with particular number combinations, such as 2, 5 and 10.

2 Practical activity. Students may need guidance to ensure that they choose a simple design. They may like to trace over an existing picture.

UNIT 4: Topic 2

Practice

1 a 9 + 5 = 14
 b 8 – 3 = 5

2 Student draws appropriate pictures and writes:
 a 14 – 6 = 8
 b 7 + 8 = 15

Challenge

1 8 – 3 = 5

2 $60 + $25 = $85

3 Teacher to check diagram. Look for students who understand they need to represent three monsters, each with three heads. Answer is 9.

Mastery

1 $2.20 + $2.35 = $4.55,
 $5 – $4.55 = 45c

2 Teacher to check. Students could be encouraged to make up word problems for their classmates to solve.

3 Teacher to check word problem. Look for students who recognise that there are five cars with four wheels on each. Answer is 20 wheels.

4 Teacher to check word problem. Look for students who recognise that there are six cookies with three chocolate chips on each. Answer is 18 chocolate chips.

UNIT 5: Topic 1

Practice

1 a Look for answers that show students' ability to choose an appropriate unit to measure the length of the items and to accurately measure using their chosen unit.
 b Teacher to check. Look for students who understand that larger units of length are more suitable for measuring longer objects and vice versa.

2 a–c Look for answers that show students' ability to choose an appropriate unit to find the area of the items and to accurately measure using their chosen unit.

Challenge

1 Teacher to check. Likely responses:
 a 4 finger widths
 b 2 finger lengths

2 a 12 squares
 b 16 squares
 c 8 squares
 d 13 squares
 e 16 squares
 f 16 squares

3 Shapes b, e and f should be circled.

Mastery

1 Teacher to check. Students could be asked to justify their choices of units. Units of length are likely to be strides or hand spans.

2 Each rectangle has an area of 6 squares.

3 Teachers could use this (or the preceding task) as an opportunity to discuss the need for standard units of measurement.
 a Each robot has an area of 17 squares.
 b Rick, Rob, Ron
 c This question could be used as a basis for a discussion on why standardised units are used in measurement in the real world. Answers may vary, e.g. "They have different areas because the squares were different sizes."

UNIT 5: Topic 2

Practice

1 Students could be asked to justify their answers.

Your hand	A Year 2 person	A table	The classroom	A pencil
Less than 1 m	About 1 m	About 1 m	More than 1 m	Less than 1 m

2 Answers may vary but should be appropriate to the object. Teacher to decide on level of accuracy expected.

3 a 4 cm b 9 cm

Challenge

1 a 13 cm
  b 10 cm
  c 12 cm

2 a m b cm
  c m d cm

3 a 30 cm b 40 m
  c 100 m d 15 cm

Mastery

1 a $9\frac{1}{2}$ cm b 6 cm
  c $4\frac{1}{2}$ cm

2 a 2 m b $1\frac{1}{2}$ m c 1 m
  d $\frac{1}{2}$ m e $2\frac{1}{2}$ m

3–4 Teacher to check and to decide on level of accuracy that is expected.

UNIT 5: Topic 3

Practice

1 A: 4 cubes, B: 10 cubes, C: 9 cubes, D: 10 cubes, E: 6 cubes, F: 8 cubes, G: 9 cubes, H: 8 cubes

2 Students could be asked to justify responses other than the following:
A, C, D, B

Challenge

1 a Student draws a square around object E.
  b Student circles objects A, C and F.
  c Student ticks object D.
  d 9 cubes

2 Likely answers are:
A: 2, B: 6, C: 3, D: 4, E: 1, F: 5

Mastery

1 a Students could carry out this activity as a cooperative group task. Likely responses are:
A: 20 cups, B: 10 cups, C: 2 cups, D: 5 cups, E: 8 cups, F: 13 cups
  b Practical activity. Teachers could ask students to share their findings with each other.

2 a 12 cubes
  b Teacher to check. Accuracy in the sketches is less important than the ability to see that a volume of 12 cubes can be represented in various orientations. Students could share their ideas with each other.

UNIT 5: Topic 4

Practice

1 A: lighter B: heavier
  C: heavier D: lighter
  E: the same mass F: heavier

2 Students could be asked to justify their responses. Likely order of mass is:
B, D, E, C, A

Challenge

1 Answers may vary. Students could share their responses with each other, discussing the reasons for their choices, e.g. "The pencil case might be almost empty, so it would be lighter than the book."

2–3 Answers may vary. This could be used as a springboard for a discussion on hefting and to what extent we can rely on it when estimating mass.

Mastery

1–5 Practical activities. Students could extend the range of weighing activities found on the page according to their skill and motivation levels. Look for students' ability to use a balance scale correctly, to identify appropriate units of mass and to accurately compare measured masses. Students could give a report on their findings and discuss their work with each other.

UNIT 5: Topic 5

Practice

1 a

b

c

2 Teacher to decide on level of accuracy required for the hour hand.
a

b

c
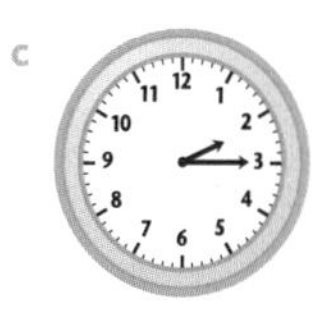

3 Teacher to decide on level of accuracy required for the hour hand.
a

b

c

4 a half past 10
  b quarter to 12
  c quarter past 4

Challenge

1 Teacher to check that the activity is appropriate to the time.
a

b

c
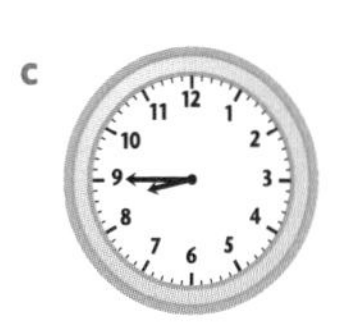
d

e
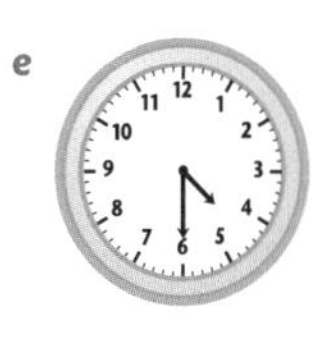
f

2 a–c Answers will vary. Teacher to check that the clock matches the written time and that the activity is appropriate to the time chosen.

Mastery

1 a 

b 

quarter to 6, 5:45 half past 8, 8:30

c 

d 

quarter past 4, 4:15 half past 6, 6:30

e 

f 

quarter to 4, 3:45 quarter past 5, 5:15

2 a Movie starts: b Movie ends:

1:45, quarter to 2 3:15, quarter past 3

UNIT 5: Topic 6

Practice

1 a 7
   b 4
   c February, 28 days
   d 5
   e 138 days

Challenge

1 a No, because it is a Saturday. Students could be asked to justify any alternative response.
   b Wednesday
   c Wednesday, 2 May
   d April, June, September and November

2 Teacher to check that the day and date match.

3 a No, because this month has 31 days and June only has 30.
   b Tuesday, Wednesday and Thursday

Mastery

1 a On the other calendar it was on a Thursday and in 2020 it is on a Saturday.
   b There are 29 days in February 2020 but only 28 on the other calendar.

2 a–e Practical activity. Students could share their responses with each other. They could also be asked to write a question for others to answer based on the information in their birthday month.

UNIT 6: Topic 1

Practice

1

2

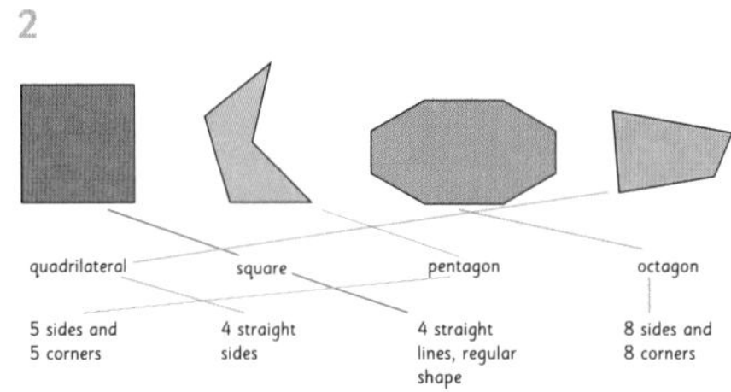

Challenge

1 Teacher to decide on acceptable level of accuracy but the drawing should be an appropriate representation of the features of the shape. Student draws:
   a a hexagon
   b a semi-circular shape
   c a square
   d a rectangle (A rhombus, a square or a parallelogram could also be drawn.)

   Students' choice of "regular" or "irregular" should match their drawing.

2 This could become part of a group discussion. Look for students who focus on the attributes of the shapes, rather than their colour or size. Possible answers are:
   a All three shapes have straight sides. They are all regular.
   b Shape A has one more side than shape B and shape B has one more side than shape C.

Mastery

1 Students might choose to focus on the different shapes that have been used, e.g, "I can see circles, ovals, triangles, semi-circles, squares, a rectangle, a kite, an octagon and a trapezium."
   Others might note the number of each type of shape: "I can see 8 triangles, 3 semi-circles, 4 circles, 3 ovals, 2 squares, a rectangle, a kite, an octagon and a trapezium."
   Other ways of describing the picture are possible and students could be encouraged to share their ideas with their peers.

2 Teachers may wish to discuss possibilities with students, ensuring they draw something that can be described using basic geometric language.

UNIT 6: Topic 2

Practice

1 sphere
   a 1 face b 0 edges
   c 0 corners

2 triangular prism
   a 5 faces b 9 edges
   c 6 corners

3 square pyramid
   a 5 faces b 8 edges
   c 5 corners

4 Student shades the cone.

Challenge

1

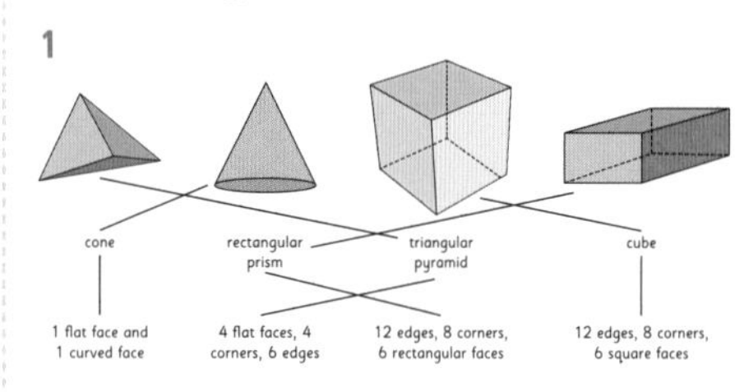

2 a triangular pyramid
   b cylinder
   c square pyramid
   b cube

Mastery

1 a 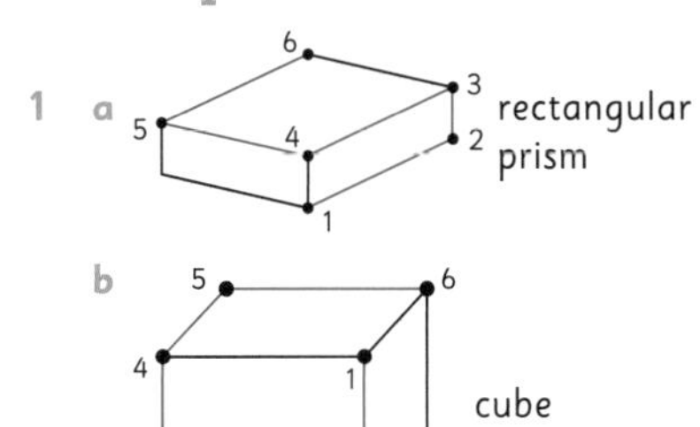

b

cube

2 a–d Student makes a reasonable attempt to draw the rectangular prism. Teachers should make available spare paper for students to use. This trial and error stage is important for working towards proficiency and teachers may wish to make students aware that much practice is usually needed before success can be achieved in 3D drawing. Some students may like to share their successful drawing strategies with their peers.

3 See also notes for question 2.

a 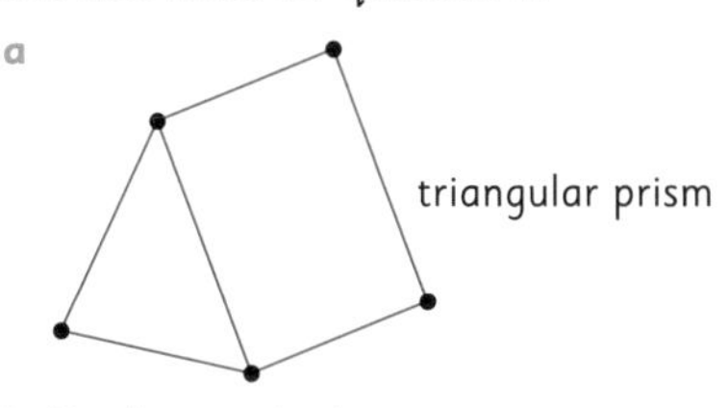

b Teacher to check.

UNIT 7: Topic 1

Practice

1 Some answers may vary. Teachers to check that answers describe the positions accurately and that appropriate vocabulary has been used. Possible answers are:

a on the left on top shelf

b the robot

c on the left on the bottom shelf/below the robot; next to the boat

d on the middle shelf on the right/above the ball; below the teddy

2 Some answers may vary. Teachers to check that the answers describe the positions accurately and that appropriate vocabulary has been used. Possible answers are:

a the table

b the window

c in the corner next to the bed

d Student describes the position of the rug appropriately.

Challenge

1 Answers will vary. Look for students who describe the positions accurately and who use appropriate vocabulary.

2 Teacher to check. Look for drawings that match instructions.

Mastery

1 a in the 3rd column on the 2nd row

b & c 

2 a–g, i

Y	O	U	
	C	A	N
R	E	A	D
T	H	I	S

h 3rd column on the bottom row

3 a

	I	
	A	M
O	K	

b Teacher to check. Possible instructions are:

I on the top row

A in the 2nd column on the 2nd row

M in the 3rd column on the 2nd row

O in the 1st column on the 3rd row

K in the 2nd column on the 3rd row.

UNIT 7: Topic 2

Practice

1 a half turn

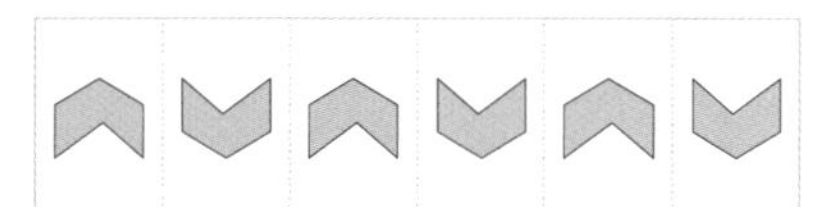

b quarter turn

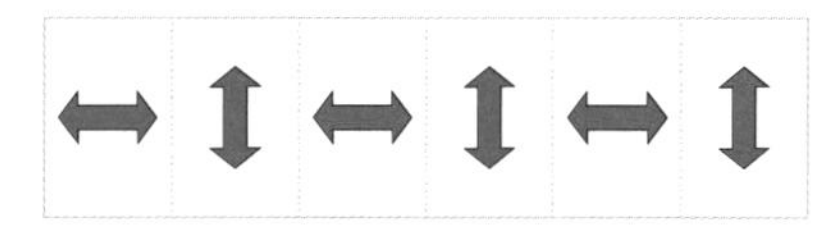

2 a

b

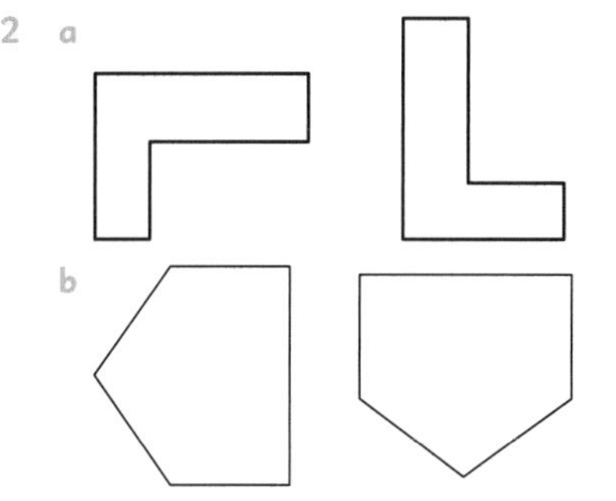

Challenge

1 a quarter turn

b

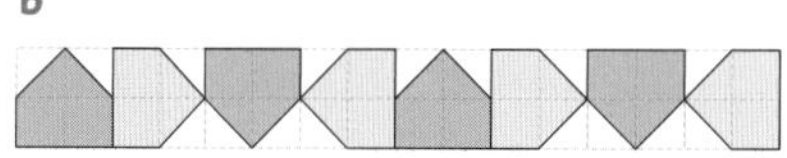

2 a half turn

b

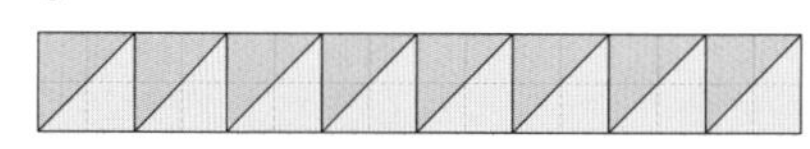

3 Turn A quarter turn, Turn B three-quarter turn, Turn C half turn, Turn D full turn

Mastery

1 a–d Teacher to check. Look for students who demonstrate understanding of the difference between the different types of turns and who can recognise how they are used in everyday situations.

UNIT 8: Topic 1

Practice

1 a

Fruit	Tick for each student	Total number
Apple	✓✓✓✓✓✓✓✓✓✓	10
Orange	✓✓✓✓✓✓✓✓	8
Strawberry	✓✓✓✓✓✓	6
Banana	✓✓✓✓✓✓✓✓✓✓✓✓	12
Other fruit	✓✓✓✓✓✓✓✓✓✓	10

b What is your favourite fruit?

c 46

Challenge

1 a–b Look for students who write a closed question, such as "Do you have a pet?" Teachers may wish to encourage students to tally the results before writing the totals.

2 a Look for students who write an appropriate open question, such as "What is your favourite snack food?"

b Teacher to check the appropriateness of the data in the table.

c Sentences should reflect information that can be acquired by reading the data.

Mastery

1 Teachers may wish to discuss the topic of closed and open questions with students before they begin in order to give more examples than are shown on the student page. Students could share their responses with each other. They could also be asked to show them to the teacher before proceeding with question 2.

2 This could be carried out as a cooperative group task in order to minimise disruption to the class.

3–4 Students could share their responses with a partner and be asked to reflect on the appropriateness of one or both of their questions in order to be able to collect data.

UNIT 8: Topic 2

Practice

1

Meal	Breakfast	Brunch	Lunch	Dinner
Tally	||	|	||||	~~||||~~
Total	2	1	4	5

2 a–e Teacher to check. Look for answers that show students' ability to choose appropriate variables that are likely to appeal to the classmates being surveyed – for example, sport, art etc. Look for students who can record the results in the correct section of the table and read the results accurately.

Challenge

1 **a** Teachers may wish to discuss the variables with students before they choose, e.g. the faces could be sorted by colour or by happy/sad/with a hat/without a hat.

b There are 54 faces:

	Happy with hat	Happy with no hat	Sad with hat	Sad with no hat	Total
Pink	7	3	0	1	**11**
Green	4	2	3	3	**12**
Blue	6	6	2	4	**18**
Yellow	3	2	4	4	**13**
Total	**20**	**13**	**9**	**12**	**54**

c Look for students whose responses accurately reflect information found in the data.

d Look for students whose questions ask for information found in the data.

Mastery

1–2 Teacher to check. For those who choose to conduct their own survey, it may be advisable for them to share their ideas and intentions before proceeding with the actual data collection. For the monster data, look for students' ability to choose appropriate variables, such as the number of eyes, the colour, the number of legs, shape. Look for students who can create a table based on the activities in this topic and who record and read the results accurately.

UNIT 8: Topic 3

Practice

1 a 10

b bird

c 5

d Answers will vary.

2 a long: 9, short: 7, medium: 8

b

9			
8			
7			
6			
5			
4			
3			
2			
1			
	Long	**Short**	**Medium**

c long

Challenge

1 Practical activity. Students should be able to complete the survey without disturbing each other.

a Answers may vary depending on what is meant by medium length, etc.

b–c Look for responses showing that the student is able to give information by interpreting the data.

2 Teachers may wish to discuss the idea of a key being used in a graph instead of numbering the vertical axis.

a play handball

b 25

c running around and football

Mastery

1 a–f Teachers to decide how much intervention will be necessary for students in these tasks. Students who are exhibiting a mastery of this topic could well be able to carry out this activity with little or no intervention from the teacher. It deals with designing a survey, followed by collecting, displaying and interpreting the data.

UNIT 9: Topic 1

Practice

1 Students may be able to justify other responses.

a certain

b unlikely

c impossible

d likely

2 a–b Answers will vary. Teacher to check. Students could share their responses with each other.

Challenge

1 **a–d**

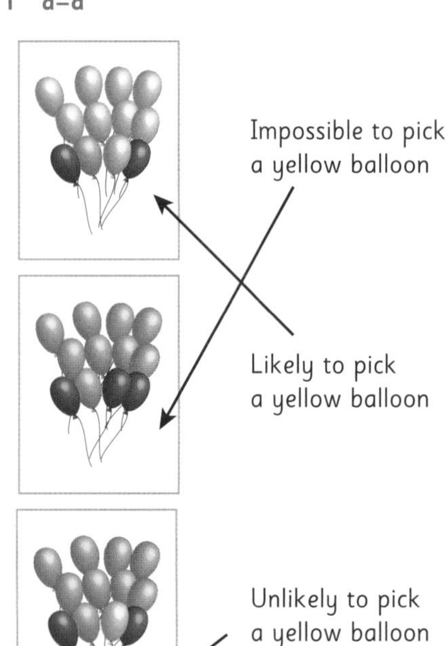

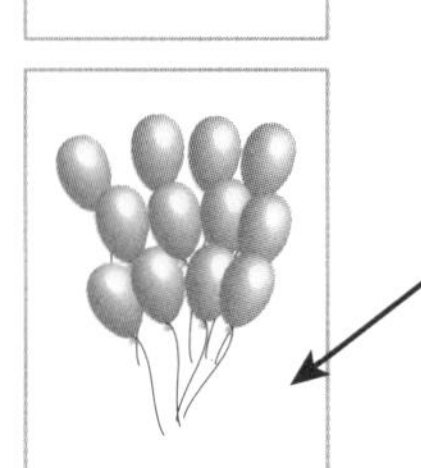

Unlikely to pick a yellow balloon

Certain to pick a yellow balloon

2 **a** unlikely **b** certain
c impossible **d** likely

3 Teacher to check appropriateness of the response.

Mastery

1 Answers may vary but should reflect the chance situations. Probable answers are:
   a 7 or more red beads and 3 or less yellow
   b 10 red beads and no yellow
   c 10 yellow beads and no red
   d 7 or more yellow beads and 3 or less red

2 Students who are working towards mastery will recognise that there are often more than four possibilities in chance situations. Vocabulary such as *equal chance*, *highly unlikely* and *less likely* could be introduced. This could be a shared activity. Possible responses are:
   a The majority of beads are red and there are only a few each of blue and yellow.
   b i–iii The majority of beads are blue. The remainder are red and yellow but there are less yellow than there are red.
   c i–ii 10 blue and 10 yellow

WORKING SPACE